KU-628-614

Penguin Books

Mathematics for Technology: 2
John Dobinson

Mathematics
for Technology: 2

John Dobinson

Penguin Books

Penguin Books Ltd, Harmondsworth, Middlesex, England
Penguin Books, 625 Madison Avenue, New York, New York 10022, U.S.A.
Penguin Books Australia Ltd, Ringwood, Victoria, Australia
Penguin Books Canada Ltd, 2801 John Street, Markham, Ontario, Canada L3R 1B4
Penguin Books (N.Z.) Ltd, 182–190 Wairau Road, Auckland 10, New Zealand

First published as *O.N. Mathemathics: 2* 1969
Reprinted 1972, 1974, 1975, 1977, 1979, 1981

Filmset in Monophoto Times by
Oliver Burridge Filmsetting Ltd, Crawley, England,
and made and printed by
Hazell Watson & Viney Ltd, Aylesbury, Bucks

Contents

Preface

This is the second of the two books planned to cover the Ordinary National Certificate and Ordinary National Diploma courses in Mathematics. Volume 1 covers a suitable first year syllabus and volume 2 the syllabus of the final year.

In each book the chapters are in the order in which it is suggested the various sections be worked. It is an order which is designed to add interest and variety to the course but there is nothing rigid about it. Indeed with the variations in syllabuses of different colleges, students are not expected to study every topic in the two volumes. Some teachers, and students, may prefer to complete the different sections, algebra, trigonometry, calculus, etc., in turn but, as the report of the Mathematical Association on the teaching of mathematics in technical colleges points out, the success of the teacher of part-time students depends to a great extent on his ability to provide links between the different branches of mathematics in each class period. A teacher would be advised to share his allotted weekly time between two or more sections of the course and so keep his students doing weekly exercises in more than one branch of mathematics.

Each topic, as far as possible, is broken up into small parts, with its set of fairly straightforward exercises, so that a small amount of theory can be followed by the student working examples for himself. At the end of each chapter are miscellaneous exercises, mostly from actual examination papers, to test the work of the whole chapter. Units are in the metric system (S.I.).

Thanks are due to the following examining bodies who have kindly allowed me to use questions from their examination papers: the East Midland Educational Union (denoted by *E.M.E.U.*), the Northern Counties Technical Examinations Council (*N.C.T.E.C.*), the Union of Educational Institutions (*U.E.I.*), the Union of Lancashire and Cheshire Institutes (*U.L.C.I.*) and the Director, Naval Education Service for H.M. Dockyard Technical Colleges (*D.T.C.*).

John Dobinson
1969

1 Theory of quadratics and partial fractions

1a Quadratic equations

A quadratic *expression* in x is an expression of the form ax^2+bx+c where a, b, c are all constants. A quadratic *equation* in x is of the form $ax^2+bx+c = 0$.

A quadratic equation has two solutions or roots and methods of obtaining these are dealt with in volume 1, chapter 2.

The equation $ax^2+bx+c = 0$ is generally used to represent any quadratic equation.

The solution by formula of $ax^2+bx+c = 0$ is easily shown by the method of completing the square to be

$$x = \frac{-b\pm\sqrt{(b^2-4ac)}}{2a}$$

The nature of the roots depends upon the value of the term $\sqrt{(b^2-4ac)}$. If b^2-4ac is negative we say that $\sqrt{(b^2-4ac)}$ is imaginary and that $\dfrac{-b\pm\sqrt{(b^2-4ac)}}{2a}$ is a complex number. Complex numbers are dealt with in a later chapter.

The expression b^2-4ac is known as the *discriminant* of the quadratic equation $ax^2+bx+c = 0$.

Both roots will be
i real if b^2-4ac is positive or zero
ii complex if b^2-4ac is negative
iii equal if b^2-4ac is zero
iv rational if b^2-4ac is a perfect square.

Example 1 For what value of k will the roots of $x^2-10x = k$ be equal? Find their values in this case.

The roots are equal if $b^2 = 4ac$.
Here $a = 1, b = -10, c = -k$.
Thus $\qquad (-10)^2 = 4\times 1\times(-k)$
$$k = -25$$

Roots are $-\dfrac{b}{2a} = \dfrac{10}{2}$

$$= 5$$

If α and β are the roots of the general quadratic $ax^2 + bx + c = 0$

dividing by a gives $\qquad x^2 + \dfrac{b}{a}x + \dfrac{c}{a} = 0$

which must be identical with $\quad (x - \alpha)(x - \beta) = 0$

i.e. $\qquad\qquad\qquad\qquad x^2 - (\alpha + \beta)x + \alpha\beta = 0$

Comparing the two equations,

$$\alpha + \beta = -\dfrac{b}{a}$$

and $\quad \alpha\beta = \dfrac{c}{a}$

Thus for a quadratic equation

i $\quad$ sum of roots $= \alpha + \beta = -\dfrac{b}{a} = -\dfrac{\text{coefficient of } x}{\text{coefficient of } x^2}$

ii product of roots $= \alpha\beta = \dfrac{c}{a} = \dfrac{\text{constant term}}{\text{coefficient of } x^2}$

These results may also be obtained by writing

$$\alpha = \dfrac{-b + \sqrt{(b^2 - 4ac)}}{2a} \text{ and } \beta = \dfrac{-b - \sqrt{(b^2 - 4ac)}}{2a}$$

and evaluating $\alpha + \beta$ and $\alpha\beta$ directly.

Example 2 Without solving the equations examine the nature of the roots and find the sum of the roots of the quadratics
i $x^2 + x + 1 = 0$ and ii $3x^2 - x - 2 = 0$

i $b^2 - 4ac = 1 - 4 \times 1 \times 1 = -3$
Thus the roots are complex.

The sum of the roots $= -\dfrac{b}{a} = -\dfrac{1}{1} = -1$

ii $b^2 - 4ac = 1 - 4 \times 3(-2) = 25$
The roots are real and rational.
Sum of roots $= \frac{1}{3}$.

Equations with given roots

Although it may be difficult to find the roots of a given equation, it is easy to find an equation when the roots are given.

Example 3 Find the equation whose roots are 2 and $-\frac{1}{2}$.

We have to find an equation which is satisfied when $x = 2$ and when $x = -\frac{1}{2}$ i.e. when $x - 2 = 0$ and when $x + \frac{1}{2} = 0$.

The required equation must be $(x-2)(x+\frac{1}{2}) = 0$

i.e. $$x^2 - 1\frac{1}{2}x - 1 = 0$$

Example 4 Find an equation whose roots are 2, 3 and -1.

In this case the equation is the cubic equation

$$(x-2)(x-3)(x+1) = 0$$
i.e. $x^3 - 4x^2 + x + 6 = 0$

1b Equations of the same form as quadratic equations

We will consider some special forms of equations of higher degree than the second.

Example 1 Solve the equations

i $x^4 - 4x^2 + 3 = 0$ ii $x(x+1) + \dfrac{12}{x(x+1)} = 8$

i Let $u = x^2$. Then the given equation becomes

$u^2 - 4u + 3 = 0$
i.e. $(u-1)(u-3) = 0$
i.e. $x^2 - 1 = 0$ or $x^2 - 3 = 0$
i.e. $x = \pm 1$ or $x = \pm\sqrt{3}$
ii Let $u = x(x+1)$. Then the given equation becomes

$$u + \frac{12}{u} = 8$$

i.e. $u^2 - 8u + 12 = 0$
$(u-2)(u-6) = 0$
Hence $x(x+1) = 2$ or $x(x+1) = 6$
$x^2 + x - 2 = 0$ or $x^2 + x - 6 = 0$
$(x+2)(x-1) = 0$ or $(x+3)(x-2) = 0$

The roots of the equation are then -2, 1, -3 and 2.

Example 2 Solve the equation $4^x - 7 \times 2^x + 10 = 0$.

As the unknown x is in the index this is called an *indicial* equation. Some indicial equations have been dealt with in volume 1, section 1e, but in this example we do not take logarithms immediately. This equation is of the form
$am^{2x} + bm^x + c = 0$

i.e. a quadratic in m^x, namely $\quad a(m^x)^2 + b(m^x) + c = 0$

To solve $\qquad\qquad\qquad\qquad\qquad 4^x - 7 \times 2^x + 10 = 0$

we write it as $\qquad\qquad\qquad\qquad 2^{2x} - 7 \times 2^x + 10 = 0$

Let $2^x = u$.

The equation becomes $\quad u^2 - 7u + 10 = 0$

$\qquad\qquad\qquad\qquad\qquad (u-2)(u-5) = 0$

Hence $\quad 2^x = 2 \quad$ or $\quad 2^x = 5$

If $\qquad 2^x = 2 \quad$ then $\quad x = 1$

If $\qquad 2^x = 5 \quad$ we take logarithms to base 10.

$x \log 2 = \log 5$

$$x = \frac{\log 5}{\log 2} = \frac{0 \cdot 6990}{0 \cdot 3010} = 2 \cdot 32$$

Thus the solutions are $x = 1$ and $x = 2 \cdot 32$.

Example 3   Solve $2 \times 9^x - 3^{2+x} + 9 = 0$.

This may be rearranged to give $\quad 2 \times 3^{2x} - 3^2 \times 3^x + 9 = 0$

i.e. $\qquad\qquad\qquad\qquad\qquad\qquad 2(3^x)^2 - 9 \times 3^x + 9 = 0$

This is obviously a quadratic equation in 3^x.

Let $u = 3^x$.

The equation becomes $\quad 2u^2 - 9u + 9 = 0$

$\qquad\qquad\qquad\qquad\qquad (2u-3)(u-3) = 0$

$u = 1 \cdot 5 \quad$ or $\quad u = 3$

$3^x = 1 \cdot 5 \quad$ or $\quad 3^x = 3$

If $3^x = 3 \quad$ then $x = 1$

If $3^x = 1 \cdot 5 \quad$ then by taking logarithms $x \log 3 = \log 1 \cdot 5$

$$x = \frac{\log 1 \cdot 5}{\log 3} = \frac{0 \cdot 1761}{0 \cdot 4771} = 0 \cdot 369$$

Solutions: $x = 1$ and $x = 0 \cdot 369$.

Exercise 1b

1   Without solving the equations, find the sums and the products of the roots of each of the following equations

(a) $x^2 + 3x - 1 = 0$ (b) $x^2 + x + 4 = 0$

(c) $3x^2 + 2x - 1 = 0$

     State whether the roots are real or complex in each of the above.

2   Find the value of k for which the equation $9x^2 - 12x + k = 7$ has equal roots.

3   Form the equations whose roots are

(a) $3, -1$ (b) $-\frac{1}{2}, 1$ (c) $k, 2k$

4 Solve the equation $x^2+(k-3)x+k = 0$, given that the roots are equal.

5 If α and β are the roots of the equation $2x^2+5x+4 = 0$ find the values of $\alpha+\beta$, $\alpha\beta$ and $\alpha^2+\beta^2$.

Find the quadratic equation with roots $\dfrac{\alpha}{\beta}$ and $\dfrac{\beta}{\alpha}$.

6 Show that the roots of the equation $4x^2-(a-2)x-1 = 0$ are real for all real values of a.

7 Solve the equations (a) $x^4-5x^2+4 = 0$ (b) $\dfrac{1}{x^4}+9 = \dfrac{10}{x^2}$

8 Solve (a) $(x^2+x)^2+3(x^2+x)-10 = 0$ (b) $2x^4-x^3+x^2-x+2 = 0$

9 Solve the indicial equations

(a) $2^x+4^{x-1} = 3$ (b) $9^x-4\times3^x+3 = 0$
(c) $7^{x-1} = 3^{2x-1}$

10 Find the greatest value of c for which the roots of the equation

$2x^2+3x+4+c = 0$

are real.

1c Partial fractions

It is often necessary to break up a single fraction into simpler fractions which are then known as *partial fractions*. A *proper* fraction is a fraction in which the numerator is of lower degree than the denominator; otherwise it is an *improper* fraction.

We consider first proper fractions where the denominator contains factors of the first degree only.

Example 1   Resolve $\dfrac{3x+7}{(x-1)(x-2)}$ into partial fractions.

Assume $\dfrac{3x-7}{(x-1)(x-2)} \equiv \dfrac{A}{x-1}+\dfrac{B}{x-2}$

where A and B are constants to be determined and the sign $\equiv$ means identically equal to.

$\dfrac{3x+7}{(x-1)(x-2)} \equiv \dfrac{A(x-2)+B(x-1)}{(x-1)(x-2)}$

15   Partial fractions

This is an identity true for all values of x. Equating numerators we have

$$3x+7 \equiv A(x-2)+B(x-1) \qquad (1.1)$$

There are two main ways of finding A and B from (1.1)

i Equating the coefficients of x on each side and the constants on each side, we have

$$3 = A+B$$
and $\quad 7 = -2A-B$
giving $\quad A = -10 \quad$ and $\quad B = 13$

ii Since equation (1.1) is an identity, it is true for all values of x.

Put $x = 1 \quad$ then $3 \times 1+7 = A(1-2)+0$
$$A = -10$$
Put $x = 2 \quad$ then $3 \times 2+7 = 0+B(2-1)$
$$B = 13$$

The student should use whichever method he prefers, and a combination of the two methods if necessary.

Thus $\quad \dfrac{3x+7}{(x-1)(x-2)} \equiv \dfrac{13}{x-2} - \dfrac{10}{x-1}$

When factors in the denominator are repeated, the partial fractions may be written as follows:

Example 2 Resolve $\dfrac{3x^2-x+4}{(x-1)^3}$ into partial fractions.

Let $\quad \dfrac{3x^2-x+4}{(x-1)^3} \equiv \dfrac{A}{x-1} + \dfrac{B}{(x-1)^2} + \dfrac{C}{(x-1)^3}$

Putting the right hand side over the common denominator $(x-1)^3$ we have

$$\dfrac{3x^2-x+4}{(x-1)^3} \equiv \dfrac{A(x-1)^2+B(x-1)+C}{(x-1)^3}$$

Equating numerators

$$3x^2-x+4 \equiv A(x-1)^2+B(x-1)+C$$

Let $x = 1 \qquad\qquad 3-1+4 = C$
$$C = 6$$
Equate coefficients of $x^2 \quad 3 = A$
Equate coefficients of $x \quad -1 = -2A+B$
$$B = -1+2A = 5$$

$$\dfrac{3x^2-x+4}{(x-1)^3} \equiv \dfrac{3}{x-1} + \dfrac{5}{(x-1)^2} + \dfrac{6}{(x-1)^3}$$

Example 3   Resolve $\dfrac{2x+5}{(x-1)^3(x-3)}$ into partial fractions.

Let $\quad \dfrac{2x+5}{(x-1)^3(x-3)} \equiv \dfrac{A}{x-1}+\dfrac{B}{(x-1)^2}+\dfrac{C}{(x-1)^3}+\dfrac{D}{x-3}$

$$\equiv \dfrac{A(x-1)^2(x-3)+B(x-1)(x-3)+C(x-3)+D(x-1)^3}{(x-1)^3(x-3)}$$

Hence $\quad 2x+5 \equiv A(x-1)^2(x-3)+B(x-1)(x-3)+C(x-3)+D(x-1)^3$

Put $x = 1 \qquad 7 = C(1-3)$

$$C = -\frac{7}{2}$$

Put $x = 3 \qquad 11 = D(3-1)^3$

$$D = \frac{11}{8}$$

Equate coefficients of x^3

$$0 = A+D$$

$$A = -\frac{11}{8}$$

Equate constants or put $x = 0$

$$5 = -3A+3B-3C-D$$

$$= \frac{33}{8}+3B+\frac{21}{2}-\frac{11}{8}$$

$$3B = 5-\frac{22}{8}-\frac{21}{2} = -\frac{33}{4}$$

$$B = -\frac{11}{4}$$

Hence $\quad \dfrac{2x+5}{(x-1)^3(x-3)} \equiv \dfrac{11}{8(x-3)}-\dfrac{11}{8(x-1)}-\dfrac{11}{4(x-1)^2}-\dfrac{7}{2(x-1)^3}$

When the denominator contains a quadratic factor then its partial fraction needs a linear numerator as in the following example.

Example 4   Resolve $\dfrac{3x^2+7x+2}{(x+1)(x^2+2x+5)}$ into partial fractions.

Let $\quad \dfrac{3x^2+7x+2}{(x+1)(x^2+2x+5)} \equiv \dfrac{A}{x+1}+\dfrac{Bx+C}{x^2+2x+5}$

Note that we have assumed for the numerator of the second fraction an expression of degree one less than the degree of the denominator.

17   Partial fractions

Thus $\dfrac{3x^2+7x+2}{(x+1)(x^2+2x+5)} \equiv \dfrac{A(x^2+2x+5)+(Bx+C)(x+1)}{(x+1)(x^2+2x+5)}$

Hence $\quad 3x^2+7x+2 \equiv A(x^2+2x+5)+(Bx+C)(x+1)$

Put $x = -1 \qquad$ then $3-7+2 = A(1-2+5)+0$

$$A = -\tfrac{1}{2}$$

Equating coefficients of $x^2 \qquad 3 = A+B$

$$B = 3\tfrac{1}{2}$$

Equating constant terms $\qquad 2 = 5A+C$

$$C = 4\tfrac{1}{2}$$

Hence $\dfrac{3x^2+7x+2}{(x+1)(x^2+2x+5)} \equiv \dfrac{7x+9}{2(x^2+2x+5)} - \dfrac{1}{2(x+1)}$

If an improper fraction has to be resolved into partial fractions it must first be reduced by division until the fractional part is a proper fraction. Thus $\dfrac{x^3-4x^2+8x+5}{(x-1)(x-2)}$ is an improper fraction and by long division becomes $x-1+\dfrac{3x+7}{(x-1)(x-2)}$. The term $\dfrac{3x+7}{(x-1)(x-2)}$ may now be reduced to partial fractions as in example 1 above.

Hence $\quad \dfrac{x^3-4x^2+8x+5}{(x-1)(x-2)} \equiv x-1+\dfrac{13}{x-2}-\dfrac{10}{x-1}$

The above methods are summarized in the following table.

Case	Example	Try partial fractions of the form
Simple linear factors in denominator	$\dfrac{2x-3}{(2x+1)(3x+2)(x-1)}$	$\dfrac{A}{2x+1}+\dfrac{B}{3x+2}+\dfrac{C}{x-1}$
Repeated linear factor in denominator	$\dfrac{x^3-2x^2+x-3}{(2x-1)^3(x+2)}$	$\dfrac{A}{2x-1}+\dfrac{B}{(2x-1)^2}+\dfrac{C}{(2x-1)^3}+\dfrac{D}{x+2}$
Quadratic factor in denominator	$\dfrac{3x^2+x+2}{(2x-1)(3x^2+4x-1)}$	$\dfrac{A}{2x-1}+\dfrac{Bx+C}{3x^2+4x-1}$
Improper fraction	$\dfrac{x^3+3x^2+x-5}{(x+1)(x+3)}$	Divide denominator into numerator until remainder is of degree one less than denominator $= x-1+\dfrac{2x-2}{(x+1)(x+3)}$ $= x-1+\dfrac{A}{x+1}+\dfrac{B}{x+3}$

Exercise Ic

Resolve into partial fractions

1   $\dfrac{x-4}{(x-2)(x-3)}$

2   $\dfrac{2x-1}{(1-x)(2-x)}$

3   $\dfrac{2x+5}{(x+3)(x+2)}$

4   $\dfrac{1}{x^2+4x+3}$

5   $\dfrac{1}{(x-a)(x-b)}$

6   $\dfrac{2}{(x+1)(x+2)(x+3)}$

7   $\dfrac{4x}{x^2-1}$

8   $\dfrac{5-x}{(x-2)^2}$

9   $\dfrac{5x^2-4x+6}{(x-2)(x+1)^2}$

10   $\dfrac{4x^2-3x+1}{(x-3)(x+1)^2}$

11   $\dfrac{x^2+1}{x(x+1)^2}$

12. $\dfrac{4x+2}{(x-1)(x^2+1)}$

13   $\dfrac{x^2+2x+5}{(x+1)(x^2+2x-2)}$

14   $\dfrac{x^3+x^2-x+3}{(x-1)(x+2)}$

15   $\dfrac{x^3+4x^2+6x+1}{x(x^2+x+1)}$

16   $\dfrac{2}{x^2(x^2-1)}$

Miscellaneous exercises 1

1   (a) Solve the equation $4^x-2^x=6$.

(b) Resolve $\dfrac{4x^2-6x-2}{(x+1)^2(x-3)}$ into three partial fractions.

(c) If p and q are the roots of the equation $x^2-2x-5=0$, find the value of $\dfrac{pq}{p+q}$.

D.T.C.

2   If α and β represent the roots of the following equations, write down the values of the sum of the roots $\alpha+\beta$, and the product of the roots $\alpha\beta$ for each. In each case state whether the roots are real or complex.

(a) $x^2+3x-2=0$ (b) $x^2+2x+2=0$
(c) $4x^2-3x-1=0$

3   Find the condition that the roots of the equation $ax^2+bx+c=0$ may be real and the further condition that they may both be positive.

If α and β are the roots of $x^2-8x+m=0$ find the value of m so that $\alpha^2+\beta^2=40$.

19   **Miscellaneous exercises 1**

4 (a) Find the possible values of a if one root of the equation $ax^2 - 9x + 3(a+1) = 0$ is twice the other.

(b) If the equation $3x^2 - 8x + a + 3 = 0$ has equal roots find the value of a and the value of the roots.

5 (a) If α and β are the roots of the equation $2x^2 + 8x + 7 = 0$, give the values of $\alpha + \beta$, $\alpha\beta$ and $\alpha^2 + \beta^2$.

(b) Solve the equation $2^{2+2x} + 3 \times 2^x - 1 = 0$.

6 The roots of $2x^2 - x + 1 = 0$ are α and β.

Without solving the equation find the values of $\alpha\beta$, $3\alpha + 3\beta$ and $\alpha^2 + \beta^2$.

Form the equation whose roots are $\alpha^2 + \beta^2$ and $\alpha\beta$.

7 Solve the equations

(a) $x^4 - 6x^2 + 8 = 0$
(b) $x^2 + 1 + \dfrac{10}{x^2 + 1} = 7$

8 Solve the equations

(a) $\sqrt{(x+5)} - \sqrt{(x-2)} = 1$
(b) $2^x \times 5^{x-1} = 8^{2x+1}$
(c) $x - 4\sqrt{x} = 5$

9 Find the partial fractions of

(a) $\dfrac{3x-2}{(x-2)(x+1)}$
(b) $\dfrac{4x^2 + x + 4}{(1+2x)(1-x)^2}$

10 Express $\dfrac{8x^2 - 9x - 2}{(x-1)^2(2x+1)}$ as the sum of three partial fractions.

11 Resolve $\dfrac{3x^2 - 2x + 4}{(x-3)(x+2)^2}$ into partial fractions.

12 (a) If the roots of the equation $x^2 + k(2x+3) - 6 = 0$ differ by 4, find the values of k.

(b) Express $\dfrac{x^3 + 10x^2 - 4x + 5}{(x^2 + x - 2)(x^2 + 1)}$ as the sum of three partial fractions.

13 (a) Solve $\sqrt{(x+6)} + \sqrt{(3x-5)} = \sqrt{(8x+1)}$.

(b) Evaluate d if f the stress in a plate is given by

$$f = \frac{P}{\pi t^2}\left[\frac{4}{3}\log_e\left(\frac{D}{d}\right) + 1\right]$$

when $f = 5180$, $P = 1200$, $t = 0.5$, $D = 6$. *U.E.I.*

20 Theory of quadratics and partial fractions

14 Solve the equations, correct to three significant figures.

(a) $6e^{2x} - 7e^x + 2 = 0$ (b) $3^{x^2} = 9^{x+1}$

(c) $\log x + \log (x+1) = \log 6$

15 (a) If p and q are the roots of the equation $2x^2 + 2x + 3 = 0$, find the value of $\dfrac{1}{p} + \dfrac{1}{q}$ without solving the equation.

(b) Solve the equation $9^x - 5 \times 3^x + 4 = 0$.

(c) Resolve $\dfrac{5x^2 - 2x - 6}{(x+2)(x-1)^2}$ into three partial fractions.

16 (a) Find x if $\sqrt{(x-4)} - \sqrt{(x-9)} = 1$

(b) Find p if $3^{p^2+4} = 81^p$

(c) Find x and y if $x^2 + y^2 = 20$

and $xy = 8$

U.L.C.I.

17 When a mass m falls a distance h on to a collar at the end of a bar of length l, it produces an instantaneous extension x and an instantaneous stress f. Equating the loss of potential energy of m to the work done in stretching the bar (the strain energy) gives

$$mg(h+x) = \frac{f^2}{2E} \times \text{volume of bar}$$

where $E = \dfrac{\text{stress}}{\text{strain}} = \dfrac{f}{x/l}$ $g = $ gravitational acceleration ($9{\cdot}81$ m s^{-2})

Obtain a formula for the maximum tensile stress induced in the bar.

A mass of 60 kg falls a distance $0{\cdot}4$ m and commences to stretch a steel bar of diameter $0{\cdot}01$ m and length 1 m. If $E = 2{\cdot}1 \times 10^{11}$ N m^{-2} calculate the maximum stress induced. What does the negative root of the equation represent?

2 Numerical methods

2a Numerical methods for use with a calculating machine

When a calculating machine is available, the solution of a problem by successive approximation is often the simplest method. A machine will swiftly and accurately carry out routine calculations repeated over and over again until an answer of sufficient accuracy is obtained.

Consider the following method of obtaining square roots which basically consists of a series of guesses, each successive guess being nearer the true answer.

Example 1   To find $\sqrt{14}$ correct to five significant figures.

Let our first guess be 4. If 4 is $\sqrt{14}$ then $\dfrac{14}{4}$ must be equal to $\sqrt{14}$.

$$\frac{14}{4} = 3{\cdot}5$$

which is not $\sqrt{14}$ but it follows that a closer approximation to $\sqrt{14}$ lies between 4 and 3·5.

Let our next guess be the average, 3·75.

$$\frac{14}{3{\cdot}75} = 3{\cdot}7333 \quad \text{to 4 D, i.e. four decimal places}$$

This is nearer to our answer. Again taking the average $\dfrac{3{\cdot}75 + 3{\cdot}7333}{2} = 3{\cdot}7416$

$$\frac{14}{3{\cdot}7416} = 3{\cdot}7417$$

Thus our answer to four significant figures is 3·742.

Continuing, $\dfrac{3{\cdot}7416 + 3{\cdot}7417}{2} = 3{\cdot}741\,65$

$$\frac{14}{3{\cdot}741\,65} = 3{\cdot}741\,66$$

Thus to five significant figures $\sqrt{14} = 3{\cdot}7417$.

The above operation may be expressed as follows:

If x_0 is an approximate answer to $\sqrt{N}$ then a better approximation is

$$x_1 = \frac{1}{2}\left(x_0 + \frac{N}{x_0}\right)$$

It follows that $x_2 = \dfrac{1}{2}\left(x_1 + \dfrac{N}{x_1}\right)$ is closer

and $\qquad x_3 = \dfrac{1}{2}\left(x_2 + \dfrac{N}{x_2}\right)$ is closer still

and the process is continued until the required degree of accuracy is obtained. The above is known as an *iterative* formula and may be written

$$x_{r+1} = \frac{1}{2}\left(x_r + \frac{N}{x_r}\right)$$

Iterative methods can be used to solve equations. In volume 1, chapter 10 we found solutions of an equation by plotting the graph of the function and finding the points where it cut the x-axis. In general, if $f(x)$ changes sign from positive to negative or negative to positive between two values of x, then the graph of $f(x)$ has crossed the x-axis making $f(x) = 0$ somewhere between these two values of x. Thus there is a root of the equation between these two values of x.

Consider the equation $x^2 + x - 1 = 0$.

When $x = 0$ then $x^2 + x - 1 = -1$
When $x = 1$ then $x^2 + x - 1 = +1$

In this case the function changes sign between 0 and 1, and somewhere between 0 and 1 is a root of $x^2 + x - 1 = 0$.

When we know the value of a root approximately, it is easy to get a closer value. If the calculations can be done by machine, then approximate methods are often the quickest and most accurate.

Example 2   Solve the equation $x^2 + x - 1 = 0$ to three decimal places.

Rearranging in the form $x(x+1) = 1$ we have

$$x = \frac{1}{x+1}$$

Suppose we take $x_0 = 1$ as a first approximation,

then $\qquad x_1 = \dfrac{1}{x_0 + 1} = \dfrac{1}{2}$ is nearer

and $\qquad x_2 = \dfrac{1}{x_1 + 1} = \dfrac{2}{3}$ is closer still.

We continue the process and show the steps in the table

x	1	0·5	0·67	0·60	0·625	0·6154	0·6190	0·6177
$1 + x$	2	1·5	1·67	1·60	1·625	1·6154	1·6190	1·6177
$\dfrac{1}{1+x}$	0·5	0·67	0·60	0·625	0·6154	0·6190	0·6177	0·6182

The root is 0·618 to 3 D (three decimal places).

Since the sum of the roots is $-\dfrac{b}{a} = -1$ then the other root is

$-1-0·618 = -1·618$

It is often necessary to find in fractional form a ratio equal to, or nearly equal to, a given decimal. This problem frequently occurs in choosing suitable gear ratios for a mechanism. When a calculating machine is available, the method used in the following example rapidly gives a series of suitable approximation.

Example 3 Find a fraction that is within 0·0001 of the decimal 0·382 79.

We make use of the fact that if $\dfrac{a}{b}$ and $\dfrac{c}{d}$ are two positive fractions then the combined fraction $\dfrac{a+c}{b+d}$ lies between them in magnitude.

We start with the fractions $\frac{3}{8} = 0·375$ and $\frac{2}{5} = 0·4$.

$\frac{3}{8}$ is too small, $\frac{2}{5}$ is too large.

Then $\dfrac{3+2}{8+5}$ is nearer $= \dfrac{5}{13} = 0·384\,62$; this is too large

$\frac{5}{13}$ is too large, $\frac{3}{8}$ is too small.

$\dfrac{5+3}{13+8}$ is nearer $= \dfrac{8}{21} = 0·380\,95$, too low.

Combining $\dfrac{8}{21}$ and $\dfrac{5}{13}$, $\dfrac{8+5}{21+13}$ will be closer to 0·382 79.

$\dfrac{8+5}{21+13} = \dfrac{13}{34} = 0·382\,35$, which is too low.

Continuing in this manner

$\dfrac{13+5}{34+13} = \dfrac{18}{47} = 0·382\,98$, this is too high

$\dfrac{18+13}{47+34}$ is nearer $= \dfrac{31}{81} = 0·382\,72$, which is acceptable.

Answer: $\dfrac{31}{81}$

Exercise 2a

With a calculating machine use iterative methods to solve the following

1   Evaluate $\sqrt{10}$ to 5 S (i.e. five significant figures) starting with $x_0 = 3·0$.

2   Evaluate $\sqrt{17}$ to 5 S.

3 Find to 4 D (i.e. four decimal places) the roots of

(a) $x^2 + 2x - 1 = 0$ (b) $x^2 + 3x - 1 = 0$

4 Use the iterative formula $x_{r+1} = \dfrac{1}{3}\left(2x_r + \dfrac{N}{x_r^2}\right)$ to find the cube root of 17 to 3 D, starting with the value $x_0 = 2$.

2b   Flow diagrams

When a programme of work for a computer or a calculating machine involves a large number of calculations of the same nature, the order of the operations is set out in a sequence known as a *flow diagram* or *flow chart*. We have referred to these earlier, in volume 1, chapter 1.

Supposing solutions are required to the set of quadratic equations represented by $ax^2 - 6x - 5 = 0$ when a varies from 1 to 10. The steps might be set out as follows:

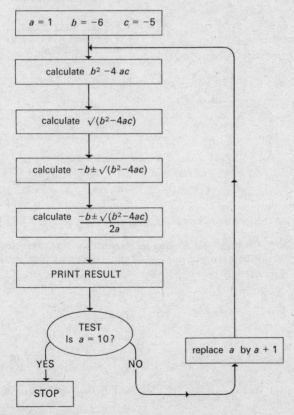

A flow diagram to solve the set of quadratic equations $ax^2 + bx - 3 = 0$ where a has integral values from 3 to 12 and b has half-integral values from $1\frac{1}{2}$ to 4, may be written as follows, where we have condensed the steps in the calculation.

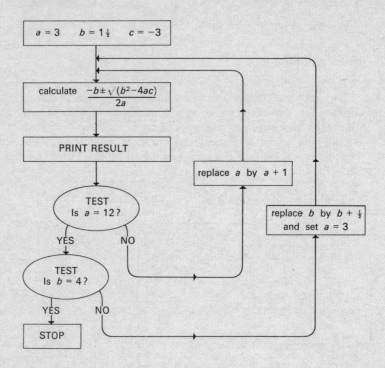

Earlier we used the iterative formula $x_{r+1} = \frac{1}{2}\left(x_r + \frac{N}{x_r}\right)$ to find a value for $\sqrt{N}$. If the acceptable error e is decided beforehand, the process is stopped when the difference between two successive approximations is less than e. A flow chart is set out as shown opposite.

Exercise 2b

1   The values of $\sqrt{\left(\dfrac{x^2 + x - 3}{2 + x}\right)}$ are required for a series of values of x from 1·3 to 2·5 in steps of 0·1, i.e. $x = 1\cdot3\,(0\cdot1)\,2\cdot5$. Draw a flow diagram for calculating these values.

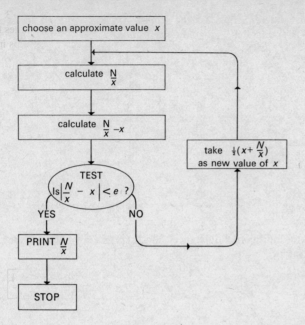

2   Draw a flow diagram to find a solution correct to three decimal places of $x^2+3x-1 = 0$ by the iterative method used in section 2a above.

3   Draw a flow diagram to solve the set of quadratic equations $2x^2+bx+c = 0$ where $b = 2\cdot2\,(0\cdot1)\,3\cdot4$ and $c = -9\,(0\cdot5)\,-4$.

2c   Finite differences

In volume 1 section 2h difference tables were built up from tabulated values of functions. For equal increments of the variable x, called the 'argument', numerical values of $f(x)$, called the 'entry', are given. We used the fact that for a polynomial of degree n the nth differences of the entry are all constant, and one could continue the table of values indefinitely.

Example 1   For the following values of $f(x)$ from $x = 0$ to $x = 10$ tabulate the finite differences for $f(x)$ up to the fourth order. Deduce that $f(x)$ is a polynomial of degree three and find it.

x	0	1	2	3	4	5	6	7	8	9	10
$f(x)$	2	−2	−4	2	22	62	128	226	362	542	772

For the first differences we use the symbol Δf (read 'delta f'), for second differences $\Delta^2 f$ (read 'delta squared f') and so on.

x	$f(x)$	Δf	$\Delta^2 f$	$\Delta^3 f$	$\Delta^4 f$
0	2				
1	-2	-4			
2	-4	-2	2		
3	2	6	8	6	0
4	22	20	14	6	0
5	62	40	20	6	0
6	128	66	26	6	0
7	226	98	32	6	0
8	362	136	38	6	0
9	542	180	44	6	0
10	772	230	50	6	

Since all the fourth order differences vanish, $f(x)$ is a polynomial of the third order. Let it be

$$ax^3 + bx^2 + cx + d$$

When $x = 0 \quad f(x) = 2$
$$d = 2$$
When $x = 1 \quad f(x) = -2$
$$a + b + c + 2 = -2$$
i.e. $\qquad a + b + c = -4$ (2.1)
When $x = 2 \quad f(x) = -4$
$$8a + 4b + 2c + 2 = -4$$
i.e. $\qquad 4a + 2b + c = -3$ (2.2)
When $x = 3 \quad f(x) = 2$
$$27a + 9b + 3c + 2 = 2$$
i.e. $\qquad 27a + 9b + 3c = 0$ (2.3)

We have to solve the simultaneous equations (2.1), (2.2) and (2.3).

From $(2.2)-(2.1)$ we get $\qquad 3a + b = 1$ (2.4)
From $(2.3)-3 \times (2.1)$ we get $\quad 24a + 6b = 12$ (2.5)
From $(2.5)-6 \times (2.4)$ we have $\qquad 6a = 6$
$$a = 1$$
Substituting in (2.4) gives $\qquad b = -2$
Substituting in (2.1) gives $\qquad c = -3$

Hence $\quad f(x) = x^3 - 2x^2 - 3x + 2$

Notation

There are three different notations in use for differences known as the forward, backward and central difference notations. We are using the forward difference notation. Thus applying suffixes f_0, f_1, f_2, etc. for the successive values of the function or entry

$$\Delta f_0 = f_1 - f_0$$
$$\Delta f_1 = f_2 - f_1$$
$$\Delta f_2 = f_3 - f_2$$

i.e. $\Delta f_r = f_{r+1} - f_r$

and for second differences

$$\Delta^2 f_0 = \Delta f_1 - \Delta f_0$$
$$\Delta^2 f_1 = \Delta f_2 - \Delta f_1$$

i.e. $\Delta^2 f_r = \Delta f_{r+1} - \Delta f_r$

and for third differences

$$\Delta^3 f_r = \Delta^2 f_{r+1} - \Delta^2 f_r$$

and so on.

Thus in example 1 above

$$f_0 = 2, \qquad f_1 = -2, \qquad f_2 = -4, \text{ etc.}$$
$$\Delta f_0 = -4, \ \Delta f_1 = -2, \ \Delta f_2 = 6, \quad \text{etc.}$$
$$\Delta^2 f_0 = 2, \quad \Delta^2 f_1 = 8, \quad \Delta^2 f_2 = 14, \quad \text{etc.}$$

Example 2   In example 1 above, verify that

i $f_2 = f_0 + 2\Delta f_0 + \Delta^2 f_0$

ii $f_3 = f_0 + 3\Delta f_0 + 3\Delta^2 f_0 + \Delta^3 f_0$

iii $f_4 = f_0 + 4\Delta f_0 + 6\Delta^2 f_0 + 4\Delta^3 f_0 + \Delta^4 f_0$

From the finite differences in example 1

$$f_0 = 2, \Delta f_0 = -4, \Delta^2 f_0 = 2, \Delta^3 f_0 = 6, \Delta^4 f_0 = 0$$

i $f_0 + 2\Delta f_0 + \Delta^2 f_0 = 2 + 2(-4) + 2$
$$= -4$$

which agrees with the given value for f_2.

ii $f_0 + 3\Delta f_0 + 3\Delta^2 f_0 + \Delta^3 f_0 = 2 + 3(-4) + 3(2) + 6$
$$= 2$$

which equals f_3.

iii $f_0 + 4\Delta f_0 + 6\Delta^2 f_0 + 4\Delta^3 f_0 + \Delta^4 f_0$
$$= 2 + 4(-4) + 6(2) + 4(6) + 0$$
$$= 22$$
$$= f_4$$

The above are examples of the Gregory–Newton interpolation formula which will be derived later.

Detection of Errors

If a table of values contains an isolated error e, this usually shows itself as an irregularity in the differences, provided the function itself is reasonably smooth. The error affects the differences according to a fan pattern as follows and so can be fairly easily detected.

Error in $f(x)$	Δf	$\Delta^2 f$	$\Delta^3 f$	$\Delta^4 f$	
				$+e$	
			$+e$	$-4e$	
	$+e$	$+e$	$-3e$	$+6e$	etc.
$+e$	$-e$	$-2e$	$+3e$	$-4e$	
		$+e$	$-e$	$+e$	

2d Interpolation

Interpolation means the determination of $f(x)$ for a value of x between two tabular values.

Suppose that we are given $f(1\cdot4) = 0\cdot323$, $f(1\cdot5) = 0\cdot341$ and we wish to estimate $f(1\cdot44)$. We can assume that the graph of $f(x)$ approximates to a straight line between $x = 1\cdot4$ and $x = 1\cdot5$ and then obtain $f(1\cdot44)$ by *linear interpolation*.

$$f(1\cdot5) - f(1\cdot4) = 0\cdot341 - 0\cdot323$$
$$= 0\cdot018 = \Delta f$$

For an increase in x of $0\cdot1$, $\Delta f = 0\cdot018$. Thus by simple proportion an increase in x of $0\cdot04$ will give an increase in $f(x)$ of

$$\frac{0\cdot04}{0\cdot1} \Delta f = \frac{0\cdot04}{0\cdot1} \times 0\cdot018$$

$$= 0\cdot0072$$

Hence $\quad f(1\cdot44) \simeq f(1\cdot4) + 0\cdot0072$
$$\simeq 0\cdot323 + 0\cdot007$$
$$\simeq 0\cdot330$$

Linear interpolation is sufficiently accurate only if the error e (see fig. 1) is sufficiently small to be neglected.

For greater accuracy it is necessary to use *higher degree interpolation* and a method of obtaining one of the many formulae is now indicated.

The Gregory–Newton interpolation formula

Suppose the numerical values of $f(x)$ are given for values where the argument x increases by equal steps h.

$$f_1 - f_0 = \Delta f_0$$
$$f_1 = f_0 + \Delta f_0$$

Figure 1

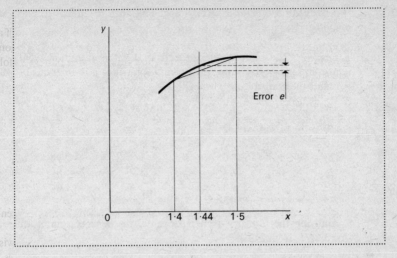

This may be written as $(1+\Delta)f_0$ where Δ is an operator acting on f_0.

$$f_2-f_1 = \Delta f_1$$
$$f_2 = f_1 + \Delta f_1 = f_0 + \Delta f_0 + \Delta f_0 + \Delta^2 f_0$$
$$= f_0 + 2\Delta f_0 + \Delta^2 f_0$$

which may be written $f_2 = (1+\Delta)^2 f_0$

$$f_3-f_2 = \Delta f_2$$
$$f_3 = f_2 + \Delta f_2$$
$$= f_0 + 2\Delta f_0 + \Delta^2 f_0 + \Delta f_0 + 2\Delta^2 f_0 + \Delta^3 f_0$$
$$= f_0 + 3\Delta f_0 + 3\Delta^2 f_0 + \Delta^3 f_0$$
$$= (1+\Delta)^3 f_0$$

Continuing in this manner f_n may be expressed in the form

$$f_n = (1+\Delta)^n f_0$$

Using the binomial theorem to expand this gives the *Gregory–Newton formula of interpolation.*

$$f_n = f_0 + n\Delta f_0 + \frac{n(n-1)}{2!}\Delta^2 f_0 + \frac{n(n-1)(n-2)}{3!}\Delta^3 f_0 + \ldots$$

Here n is a positive integer but the result still holds for fractional and negative values. Thus if we require f_p where p is a fraction of the interval h we have

$$f_p = f_0 + p\Delta f_0 + \frac{p(p-1)}{2!}\Delta^2 f_0 + \frac{p(p-1)(p-2)}{3!}\Delta^3 f_0 + \ldots$$

The right hand side is an infinite series but usually all the terms beyond a certain point become negligible or, as in the case of polynomials, they vanish.

If the second difference term is negligible then it reduces to linear interpolation.

Example 3   The following figures are from skeleton steam tables and give the temperatures t of dry saturated steam at pressures p in intervals of 5 units. Estimate the temperature when $p = 17$.

p	t	Δf	$\Delta^2 f$	$\Delta^3 f$
10	193·2			
		19·8		
15	213·0		−4·8	
		15·0		1·9
20	228·0		−2·9	
		21·1		1·0
25	240·1		−1·9	
		10·2		0·7
30	250·3		−1·2	
		9·0		
35	259·3			

Applying the Gregory–Newton formula to the line running diagonally downwards from 213·0

$f_0 = 213\cdot0,\ \Delta f_0 = 15\cdot0,\ \Delta^2 f_0 = -2\cdot9,\ \Delta^3 f_0 = 1\cdot0$

Succeeding differences are negligible.

$$f_p = f_0 + p\Delta f_0 + \frac{p(p-1)}{1\times2}\Delta^2 f_0 + \frac{p(p-1)(p-2)}{1\times2\times3}\Delta^3 f_0 + \ldots$$

Here p is the fraction of 2 units of the interval 5 units

$p = \frac{2}{5} = 0\cdot4$

$$f_{0\cdot4} = 213\cdot0 + 0\cdot4\times15\cdot0 + \frac{0\cdot4(0\cdot4-1)}{1\times2}(-2\cdot9) + \frac{0\cdot4(0\cdot4-1)(0\cdot4-2)}{1\times2\times3}(1\cdot0)$$

$$= 213\cdot0 + 6\cdot0 + 0\cdot348 + 0\cdot107$$
$$= 219\cdot5 \quad \text{and this is the required temperature.}$$

Exercise 2d

1   For the following values of $f(x)$ for $x = -2\,(1)\,6$ tabulate the finite differences up to the third order. Show that $f(x)$ is a polynomial in x and find it.

x	−2	−1	0	1	2	3	4	5	6
$f(x)$	12	3	−2	−3	0	7	18	33	52

2   Evaluate $f(x) = x^3 - 2x^2 + 2$ for $x = 0\,(0\cdot1)\,0\cdot4$. Working from the differences, tabulate $f(x)$ for $x = 0\cdot4\,(0\cdot1)\,1$ and check the final value by calculating $x^3 - 2x^2 + 2$ when $x = 1$.

3   The following table gives values of $f(x)$ for $x = 0\,(1)\,10$. Tabulate the finite

differences up to the fourth order and show that $f(x)$ is a polynomial in x. State the order of the polynomial and find it.

x	0	1	2	3	4	5	6	7	8	9	10
$f(x)$	3	-2	-5	0	19	58	123	220	355	534	763

4 If $f(0.5) = 0.1870$ and $f(0.6) = 0.1476$, by linear interpolation find $f(0.55)$ and $f(0.57)$.

5 If $\log_e 7.2 = 1.9741$ and $\log_e 7.3 = 1.9879$, find $\log_e 7.24$ and $\log_e 7.275$.

6 The following table gives the values of $f(x)$ for intervals of the argument $x = 2\ (1)\ 8$. Tabulate the differences up to the third order and using the Gregory–Newton interpolation formula estimate the values of $f(3.3)$ and $f(4.6)$.

x	2	3	4	5	6	7	8
$f(x)$	240.2	263.0	292.0	328.0	370.3	415.4	461.7

7 Using the Gregory–Newton formula estimate the values of $f(0.35)$ and $f(0.33)$ from the following table:

x	$f(x)$	Δf	$\Delta^2 f$	$\Delta^3 f$
0.3	0.3798			
0.4	0.4232	434		
0.5	0.4627	395	-39	6
0.6	0.4989	362	-33	

8 One value of the function in the following table is in error. Locate the error and correct the value of the function.

x	0	1	2	3	4	5	6	7	8	9	10
$f(x)$	1.00	1.03	1.15	1.48	2.14	3.28	4.93	7.30	10.48	14.59	19.75

Miscellaneous exercises 2

(1, 3, 5 and 12 are for solution using a calculating machine.)

1 Using iterative methods
(a) evaluate $\sqrt{19}$ to 5 S.
(b) solve the equation $x^2 - 2x - 1 = 0$ to 4 D by rearranging in the form
$$x = \frac{1}{x-2}.$$

2 Draw up a flow diagram for solving the set of quadratic equations

$$ax^2 + 2x - 3 = 0$$

when a varies from 3 to 12.

3 Find in fractional form a ratio that is within 0·0001 of 0·821 48 by starting with the fractions $\frac{4}{5}$ and $\frac{7}{8}$ and using the fact that if $\frac{a}{b}$ and $\frac{c}{d}$ are two positive fractions, then $\frac{a+c}{b+d}$ lies between them in magnitude.

4 (a) If $\sqrt[3]{41} = 3·448$ and $\sqrt[3]{42} = 3·476$, estimate by linear interpolation $\sqrt[3]{41·65}$ and $\sqrt[3]{41·83}$.
   (b) If $e^{0·73} = 2·075\,08$ and $e^{0·74} = 2·095\,94$, estimate $e^{0·7325}$ and $e^{0·7364}$.

5 Solve the linear equations

$$2·7x + 0·9y + 0·3z = 6·4$$
$$0·4x + 1·1y + 0·1z = 18·7$$
$$1·6x + 0·8y + 0·4z = 10$$

by setting out in tabular form and using a calculating machine (see volume 1, section 2g).

6 Evaluate $f(x) = 2x^3 - x + 3$ for $x = 0\,(0·1)\,0·4$.

Tabulate the finite differences and from these continue the entry for $x = 0·5\,(0·1)\,1·2$. Check the final value by calculating $2x^3 - x + 3$ when $x = 1·2$.

7 Show that the second differences are constant for the following table and hence that $f(x)$ must be of the form $ax^2 + bx + c$. Find the values of a, b and c.

x	-2	1	0	1	2	3	4	5	6
$f(x)$	15	6	1	0	3	10	20	36	55

8 The figures given are taken from steam tables for dry saturated steam. Form a difference table for h (the enthalpy) up to the third order of differences. Estimate the value of h at $p = 18$
   (a) by linear interpolation.
   (b) using the Gregory–Newton formula as far as second differences.

p	15	20	25	30
h	1150·8	1156·3	1160·6	1164·1

9   Tabulate the differences for $f(x)$ up to the third order.

x	40	50	60	70	80	90
$f(x)$	1·6776	1·6597	1·6450	1·6327	1·6219	1·6124

By interpolation, using up to third order differences, find the values of $f(42)$ and $f(53)$.

10   The following table gives the values of sec x for the given values of x.

x	45°	46°	47°	48°	49°
sec x	1·414 214	1·439 557	1·466 279	1·494 477	1·524 253

Determine the value of sec 45·8° correct to 6 S using the Gregory–Newton formula.

11   The distance s cm travelled by a point P on a straight line mechanism from a fixed point after a time t s is given in the table below.

t	0	2	4	6	8	10
s	6	−2	−2	6	22	46

By forming a difference table for s, find the equation giving s as a function of t. Calculate the velocity and acceleration of P when $t = 5$ s.

12   (a) Evaluate $\sqrt{\left(\dfrac{63·251 \times 2·6871}{15·734}\right)}$ expressing your answer to an appropriate accuracy assuming the given numbers are rounded to 5 S.

       (b) The iterative formula $x_{r+1} = x_r(2 - Nx_r)$ may be used to find the reciprocal of N without resorting to division. Use it to find the reciprocal of 8·341 to 7 D.

3 Compound angle formulae and periodic functions

3a Revision

In volume 1 the following identities were derived

$$\sin (A+B) = \sin A \cos B + \cos A \sin B \qquad (3.1)$$
$$\sin (A-B) = \sin A \cos B - \cos A \sin B \qquad (3.2)$$
$$\cos (A+B) = \cos A \cos B - \sin A \sin B \qquad (3.3)$$
$$\cos (A-B) = \cos A \cos B + \sin A \sin B \qquad (3.4)$$

From identities (3.1) and (3.3) by putting $A = B$ the double-angle formulae followed.

$$\sin 2A = 2 \sin A \cos A$$
$$\cos 2A = \cos^2 A - \sin^2 A$$
$$= 2 \cos^2 A - 1 \qquad \text{since } \sin^2 A + \cos^2 A = 1$$
$$= 1 - 2 \sin^2 A$$

Corresponding identities for tangents were

$$\tan (A+B) = \frac{\tan A + \tan B}{1 - \tan A \tan B}$$

$$\tan (A-B) = \frac{\tan A - \tan B}{1 + \tan A \tan B}$$

$$\text{and} \quad \tan 2A = \frac{2 \tan A}{1 - \tan^2 A}$$

The double-angle formulae may be used in conjunction with the cosine formula to give formulae for $\sin \frac{1}{2}A$, $\cos \frac{1}{2}A$, $\tan \frac{1}{2}A$, etc. in terms of the sides of a triangle and in a form suitable for use with logarithms.

Substituting $\quad \cos A = 2 \cos^2 \dfrac{A}{2} - 1$

in $\qquad \cos A = \dfrac{b^2 + c^2 - a^2}{2bc}$

we get
$$2\cos^2\frac{A}{2} = \frac{b^2+c^2-a^2}{2bc}+1$$

$$= \frac{b^2+c^2-a^2+2bc}{2bc}$$

$$= \frac{(b+c)^2-a^2}{2bc}$$

$$= \frac{(b+c+a)(b+c-a)}{2bc}$$

Let $\quad a+b+c = 2s$
then $\quad b+c-a = 2s-2a = 2(s-a)$

$$\cos^2\frac{A}{2} = \frac{2s\times 2(s-a)}{4bc} = \frac{s(s-a)}{bc}$$

Thus $\quad \cos\dfrac{A}{2} = \sqrt{\dfrac{s(s-a)}{bc}}$

Similarly, using $\cos A = 1-2\sin^2\frac{1}{2}A$ we get

$$\sin\frac{A}{2} = \sqrt{\frac{(s-b)(s-c)}{bc}}$$

and since $\quad \dfrac{\sin\frac{1}{2}A}{\cos\frac{1}{2}A} = \tan\frac{1}{2}A$

then $\quad \tan\dfrac{A}{2} = \sqrt{\dfrac{(s-b)(s-c)}{s(s-a)}}$

Any one of these formulae may be used to calculate the angles of a triangle when given the three sides a, b and c.

Exercise 3a Revision

1 Verify the formulae $\sin(A-B) = \sin A\cos B - \cos A\sin B$
$$\text{and } \tan(A+B) = \frac{\tan A+\tan B}{1-\tan A\tan B}$$

(a) when $A = \dfrac{\pi}{3}$ rad and $B = \dfrac{\pi}{6}$ rad (b) when $A = \dfrac{\pi}{6}$ rad and $B = \dfrac{\pi}{12}$ rad

2 If $\sin A = \frac{4}{5}$ find the values of $\sin 2A$, $\cos 2A$ and $\tan 2A$ without using tables.

3 Using the ratios of $45°$ and $30°$, find the values of $\sin 75°$, $\sin 15°$ and $\cos 15°$ without using tables and leaving your answers in surd form.

4 If $\tan A = \frac{12}{5}$, $\sin B = \frac{4}{5}$ and A and B are both acute angles, obtain the values of $\sin(A+B)$ and $\tan(A-B)$ without using tables.

5 Show that $\sin\left(\dfrac{\pi}{4}+A\right)\cos\left(\dfrac{\pi}{4}-B\right)+\cos\left(\dfrac{\pi}{4}+A\right)\sin\left(\dfrac{\pi}{4}-B\right) = \cos(A-B)$.

Prove the following

6    $\sin\left(\dfrac{\pi}{3}+x\right)-\sin\left(\dfrac{\pi}{3}-x\right)=\sin x$

7    $\cos\left(x-\dfrac{\pi}{6}\right)-\cos\left(x+\dfrac{\pi}{6}\right)=\sin x$

8    $\dfrac{\cos 2\theta}{\cos\theta+\sin\theta}=\cos\theta-\sin\theta$

9    $\dfrac{\sin 2A}{1+\cos 2A}=\tan A$

10    $\dfrac{1-\cos 2x}{1+\cos 2x}=\tan^2 x$

11   Prove that in any triangle $\tan\dfrac{A}{2}=\sqrt{\dfrac{(s-b)(s-c)}{s(s-a)}}$ where $s=\tfrac{1}{2}(a+b+c)$.
       Solve the triangle in which $a=270,\ b=372,\ c=198$.

12   Solve triangle ABC in which $a=52\cdot8,\ b=39\cdot3$ and $c=72\cdot1$.

3b To change products into sums or differences

A further set of identities which enable products of sines and cosines to be changed to sums or differences may be obtained as follows:

Adding equations (3.1) and (3.2) from section 3a above gives

$\sin(A+B)+\sin(A-B)=2\sin A\cos B$

Subtracting equation (3.2) from (3.1) we get

$\sin(A+B)-\sin(A-B)=2\cos A\sin B$

Adding (3.3) and (3.4) gives

$\cos(A+B)+\cos(A-B)=2\cos A\cos B$

Subtracting (3.3) from (3.4)

$\cos(A-B)-\cos(A+B)=2\sin A\sin B$

Their usefulness is more apparent if written in the form

$\sin A\cos B=\tfrac{1}{2}\left[\sin(\text{sum})+\sin(\text{difference})\right]$
$\cos A\sin B=\tfrac{1}{2}\left[\sin(\text{sum})-\sin(\text{difference})\right]$
$\cos A\cos B=\tfrac{1}{2}\left[\cos(\text{sum})+\cos(\text{difference})\right]$
$\sin A\sin B=\tfrac{1}{2}\left[\cos(\text{difference})-\cos(\text{sum})\right]$

Here difference = first angle − second angle.

These conversions will be found to be especially useful in the calculus when integrating products of sines and cosines.

Example 1 (a) Express as a sum or a difference

              i sin $7x$ sin $5x$ ii $2 \cos 11x \cos 3x$

      (b) Evaluate without the use of tables

              i sin $75°$ sin $15°$ ii $8 \sin 45° \cos 15°$

(a) i $\sin 7x \sin 5x = \frac{1}{2}\left[\cos(7x-5x)-\cos(7x+5x)\right]$
$$= \tfrac{1}{2}\cos 2x - \tfrac{1}{2}\cos 12x$$

   ii $2\cos 11x \cos 3x = 2\times\frac{1}{2}\left[\cos(11x+3x)+\cos(11x-3x)\right]$
$$= \cos 14x + \cos 8x$$

(b) i $\sin 75° \sin 15° = \frac{1}{2}\left[\cos(75°-15°)-\cos(75°+15°)\right]$
$$= \tfrac{1}{2}\cos 60° - \tfrac{1}{2}\cos 90°$$
$$= \tfrac{1}{2}\times\tfrac{1}{2} - \tfrac{1}{2}\times 0$$
$$= \tfrac{1}{4}$$

   ii $8 \sin 45° \cos 15° = 8\times\frac{1}{2}\left[\sin(45°+15°)+\sin(45°-15°)\right]$
$$= 4\sin 60° + 4\sin 30°$$
$$= 4\times\frac{\sqrt{3}}{2} + 4\times\frac{1}{2}$$
$$= 2\sqrt{3}+2$$

Exercise 3b

Express as a sum or difference the following

1	$2\cos 5\theta \cos 3\theta$	2	$2\sin 5\theta \sin 3\theta$
3	$2\cos 4\theta \sin 3\theta$	4	$2\sin 7\theta \cos 4\theta$
5	$2\cos 11\theta \cos 3\theta$	6	$2\sin 54° \sin 26°$
7	$\sin 25° \cos 65°$	8	$\cos 130° \cos 10°$
9	$\sin 40° \cos 70°$	10	$\sin 5x \cos x$
11	$\cos 3x \cos x$	12	$\sin 4x \sin 2x$
13	$2\cos(x+2y)\sin(x+y)$	14	$2\sin(x+y)\sin(x-y)$

Prove that

15   $\sin(x+45°)\sin(45°-x) = \frac{1}{2}\cos 2x$

16   $\sin\dfrac{x}{2}\sin\dfrac{7x}{2}+\sin\dfrac{3x}{2}\sin\dfrac{11x}{2} = \sin 2x \sin 5x$

17   $\dfrac{\sin 8x \cos x - \sin 6x \cos 3x + \sin 3x}{\cos 2x \cos x - \sin 3x \sin 4x - \cos 3x} = \tan 7x$

18   $\sin(A+B)\sin(A-B) = \sin^2 A - \sin^2 B = \cos^2 B - \cos^2 A$

39   **To change products into sums or differences**

3c   To change sums and differences into products

The product formulae derived above may be used in the converse form. Starting with the equations

$$\sin (A+B)+\sin (A-B) = 2 \sin A \cos B$$
$$\sin (A+B)-\sin (A-B) = 2 \cos A \sin B$$
$$\cos (A+B)+\cos (A-B) = 2 \cos A \cos B$$
$$\cos (A+B)-\cos (A-B) = -2 \sin A \sin B$$

let $A+B = P$ $A-B = Q$

whence $A = \dfrac{P+Q}{2}$ and $B = \dfrac{P-Q}{2}$, giving four more formulae

$$\sin P+\sin Q = 2 \sin \frac{P+Q}{2} \cos \frac{P-Q}{2}$$

$$\sin P-\sin Q = 2 \cos \frac{P+Q}{2} \sin \frac{P-Q}{2}$$

$$\cos P+\cos Q = 2 \cos \frac{P+Q}{2} \cos \frac{P-Q}{2}$$

$$\cos P-\cos Q = -2 \sin \frac{P+Q}{2} \sin \frac{P-Q}{2}$$

Notice that in the final formula since $\sin (-\theta) = -\sin \theta$

$$\sin \frac{(P-Q)}{2} \text{ may be written } -\sin \frac{Q-P}{2}$$

and the formula becomes

$$\cos P-\cos Q = 2 \sin \frac{P+Q}{2} \sin \frac{Q-P}{2}$$

It may help to memorize them if put in the form

$$\sin + \sin = 2 \sin \text{ (half sum) } \cos \text{ (half difference)}$$
$$\sin - \sin = 2 \cos \text{ (half sum) } \sin \text{ (half difference)}$$
$$\cos + \cos = 2 \cos \text{ (half sum) } \cos \text{ (half difference)}$$
$$\cos - \cos = 2 \sin \text{ (half sum) } \sin \text{ (half difference reversed)}$$

Example 1   Prove that

(a) $\dfrac{\sin 7\theta+\sin 5\theta}{\cos 7\theta-\cos 5\theta} = -\cot \theta$

(b) $\sin \theta+\sin (\theta+120°)+\sin (\theta+240°) = 0$

(a) L. H. S. $= \dfrac{\sin 7\theta + \sin 5\theta}{\cos 7\theta - \cos 5\theta}$

$= \dfrac{2 \sin \frac{1}{2}(7\theta + 5\theta) \cos \frac{1}{2}(7\theta - 5\theta)}{2 \sin \frac{1}{2}(7\theta + 5\theta) \sin \frac{1}{2}(5\theta - 7\theta)}$

$= \dfrac{\sin 6\theta \cos \theta}{\sin 6\theta \sin (-\theta)}$

$= \dfrac{\cos \theta}{-\sin \theta}$

$= -\cot \theta = $ R. H. S.

(b) L. H. S. $= \left[\sin (\theta + 240°) + \sin \theta \right] + \sin (\theta + 120°)$

$= 2 \sin \frac{1}{2}(2\theta + 240°) \cos \frac{1}{2}(240°) + \sin (\theta + 120°)$

$= 2 \sin (\theta + 120°) \cos 120° + \sin (\theta + 120°)$

$= \left[2 \cos 120° + 1 \right] \sin (\theta + 120°)$

$= \left[2(-\frac{1}{2}) + 1 \right] \sin (\theta + 120°)$

$= 0 = $ R. H. S.

Example 2 Assuming the sine rule for any triangle, prove that

$\tan \dfrac{B-C}{2} = \dfrac{b-c}{b+c} \cot \dfrac{A}{2}.$

Solve triangle ABC in which $a = 395·6$, $b = 372·5$ and $C = 0·65$ rad $= 37° 15'$.

By the sine rule $\dfrac{b}{c} = \dfrac{\sin B}{\sin C}$

$\dfrac{b-c}{b+c} = \dfrac{\sin B - \sin C}{\sin B + \sin C}$

$= \dfrac{2 \cos \frac{1}{2}(B+C) \sin \frac{1}{2}(B-C)}{2 \sin \frac{1}{2}(B+C) \cos \frac{1}{2}(B-C)}$

$= \dfrac{\tan \frac{1}{2}(B-C)}{\tan \frac{1}{2}(B+C)}$

Since $A+B+C = 180°$ $\dfrac{B+C}{2} = 90° - \dfrac{A}{2}$

$\tan \dfrac{B+C}{2} = \tan \left(90° - \dfrac{A}{2} \right) = \cot \dfrac{A}{2}$

Hence $\tan \dfrac{B-C}{2} = \dfrac{b-c}{b+c} \cot \dfrac{A}{2}$

Cyclic interchange of the letters gives corresponding formulae for $\tan \frac{1}{2}(C-A)$ and $\tan \frac{1}{2}(A-B)$

$$\tan \frac{A-B}{2} = \frac{a-b}{a+b} \cot \frac{C}{2} = \frac{395 \cdot 6 - 372 \cdot 5}{395 \cdot 6 + 372 \cdot 5} \cot 18° \ 38'.$$

$$= \frac{23 \cdot 1}{768 \cdot 1} \cot 18° \ 38' = 0 \cdot 0892$$

$$\frac{A-B}{2} = 5° \ 6'$$

$$A - B = 10° \ 12'$$
$$A + B = 180° - 37° \ 15'$$
$$\qquad = 142° \ 45'$$

Hence $A = 76° \ 29' = 1 \cdot 335$ rad

and $\quad B = 66° \ 16' = 1 \cdot 157$ rad

By the sine rule $c = \dfrac{b \sin C}{\sin B} = \dfrac{372 \cdot 5 \sin 37° \ 15'}{\sin 66° \ 16'}$

$$= 246 \cdot 2.$$

Exercise 3c

Express as products

1 $\sin 75° - \sin 15°$ 2 $\cos 75° + \cos 15°$ 3 $\cos 52° - \cos 28°$

4 $\sin 50° + \sin 10°$ 5 $\sin 32° + \sin 24°$ 6 $\cos 80° - \cos 40°$

7 $\sin 6x + \sin 4x$ 8 $\cos 6x + \cos 4x$ 9 $\sin 6x - \sin 4x$

10 $\cos 6x - \cos 4x$ 11 $\sin 7x - \sin 5x$ 12 $\cos 3x + \cos x$

Prove the following

13 $\dfrac{\sin 5\theta - \sin 3\theta}{\cos 5\theta + \cos 3\theta} = \tan \theta$ 14 $\dfrac{\sin A + \sin B}{\cos A + \cos B} = \tan \dfrac{A+B}{2}$

15 $\dfrac{\sin x + \sin 3x}{\cos x + \cos 3x} = \tan 2x$ 16 $\dfrac{\sin x + \sin 2x}{\cos x - \cos 2x} = \cot \dfrac{x}{2}$

17 $\sin 50° - \sin 70° + \sin 10° = 0$

18 $\sin \theta + \sin \left(\theta + \dfrac{2\pi}{3} \right) + \sin \left(\theta + \dfrac{4\pi}{3} \right) = 0$

19 $\dfrac{\sin A + \sin 3A + \sin 5A}{\cos A + \cos 3A + \cos 5A} = \tan 3A$

20    $\dfrac{\sin A - \sin 5A + \sin 9A - \sin 13A}{\cos A - \cos 5A - \cos 9A + \cos 13A} = \cot 4A$

21   Prove that in any triangle $\tan\dfrac{B-C}{2} = \dfrac{b-c}{b+c}\cot\dfrac{A}{2}$.

Solve the triangle in which $A = 0\!\cdot\!389$ rad, $b = 432$ and $c = 312$.

22   Solve the triangle in which $b = 84\!\cdot\!2$, $c = 61\!\cdot\!6$ and $A = 0\!\cdot\!498$ rad.

3d   Periodic functions

In volume 1, section 3i the graph of the function $y = a \sin px$ was plotted and the values of y were seen to oscillate between the peak values $+a$ and $-a$ and the graph repeated itself every time x increased by $\dfrac{2\pi}{p}$ radians; a was called the amplitude of the oscillation and $\dfrac{2\pi}{p}$ the period.

Figure 2

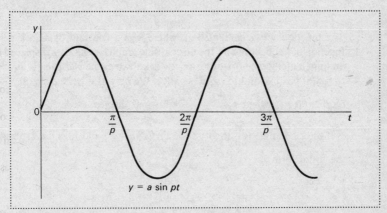

$y = a \sin pt$

If we consider the periodic function represented by $y = a \sin pt$ (see fig. 2), where t is the time in seconds, then the period $\dfrac{2\pi}{p}$ becomes the time in seconds for the motion to repeat itself. The number of repetitions or *cycles* which occur per second is known as the *frequency* and for $y = a \sin pt$ the frequency f is given by

$$f = \frac{1\ \text{s}}{(2\pi/p)\ \text{s}} = \frac{p}{2\pi}$$

and $p = 2\pi f$

Thus $y = a \sin 2\pi ft$ represents a periodic function of frequency f.

A periodic function of frequency 50 Hz may be represented by the sine function $y = a \sin 2\pi(50)t$

$$= a \sin 100\pi t$$

Figure 3

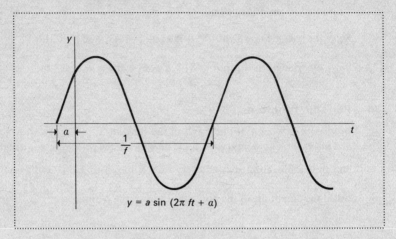

$$y = a \sin (2\pi \, ft + a)$$

The function $y = a \sin (2\pi ft + \alpha)$, where α is a constant, is also a periodic function of frequency f and the normal sine graph (see fig. 3) is displaced by an amount α along the horizontal axis; α is known as the *phase angle* and is said to *lead* if positive and to *lag* if negative. We may also write $a \sin (2\pi ft + \alpha)$ as

$a \sin 2\pi f(t + t_1)$ where $t_1 = \dfrac{\alpha}{2\pi f}$ is known as the *phase time*.

If the lead angle is $\dfrac{\pi}{2}$ then $y = a \sin\left(2\pi ft + \dfrac{\pi}{2}\right)$ and may be expressed in the cosine form as

$y = a \cos 2\pi ft$.

Example 1   The mains voltage supply has a maximum value of 240 V and a frequency of 50 Hz. Express the voltage V as a periodic function of the time t if at a time when $t = 0$ the voltage is i 0 V, ii 240 V, iii 120 V.

i   The voltage may be represented by periodic function

$V = a \sin 2\pi ft$

The frequency $f = 50$ and the amplitude $a = 240$

$V = 240 \sin 100\pi t$

ii   When $t = 0$ the voltage has its maximum value. The function starts at its peak value and has a phase angle lead of $\dfrac{\pi}{2}$; f and a are the same as in (i) hence

$$V = 240 \sin\left(100\pi t + \frac{\pi}{2}\right)$$

or in the cosine form

$$V = 240 \cos 100\pi t$$

iii Since the voltage has some intermediate value when $t = 0$ we represent it by

$V = 240 \sin(100\,\pi t + \alpha)$ where α is the phase angle.

Substituting $V = 120$ when $t = 0$ we get

$$120 = 240 \sin \alpha$$

$$\sin \alpha = \frac{1}{2} \text{ and } \alpha = \frac{\pi}{6}$$

where α must be expressed in radians.

The voltage may be expressed in the form

$$V = 240 \sin\left(100\pi t + \frac{\pi}{6}\right)$$

Example 2 Two sets of oscillations of equal amplitudes and frequencies of 26 and 22 respectively are superimposed. They can be represented as sine functions with no phase difference. Find the resultant and illustrate with a sketch.

Let the displacements y_1 and y_2 be represented by sine waves of the form
$$y = a \sin 2\pi ft$$
Then $\quad y_1 = a \sin 2\pi(26)t = a \sin 52\pi t$
and $\quad y_2 = a \sin 2\pi(22)t = a \sin 44\pi t$
where a is the common amplitude.

By addition their combined displacement y is given by

$$y = y_1 + y_2 = a(\sin 52\pi t + \sin 44\pi t)$$

$$= 2a \sin \frac{52\pi + 44\pi}{2} t \cos \frac{52\pi - 44\pi}{2} t$$

$$= 2a \sin 48\pi t \cos 4\pi t$$

This is the product of a sine wave of frequency 24 and a cosine wave of frequency 2.

We may think of this as a sine wave of frequency 24 with a varying amplitude of frequency 2, i.e. a relatively high frequency wave of slowly varying amplitude. The diagram is shown in fig. 4.

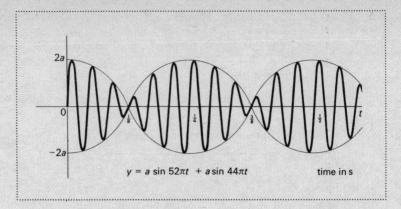

Figure 4

$$y = a \sin 52\pi t + a \sin 44\pi t$$

time in s

This example illustrates the production of 'beats' between waves of different frequencies. With sound waves this may be heard as a beating note, or, if the frequencies are close, as a note of varying intensity. In radio transmissions it is met in amplitude modulation where a carrier wave of constant frequency is made to vary in amplitude in accordance with a modulating audio frequency.

Exercise 3d

1   State the amplitudes, periods, frequencies and phase times of the following

(a) $y = 150 \sin 20t$ (b) $y = 200 \cos 100\,\pi t$

(c) $y = y_0 \sin \left(\dfrac{at}{b} + c \right)$ (d) $s = 4 \cdot 2 \sin (60t + 0 \cdot 02)$

(e) $s = 0 \cdot 17 \sin (10t - 1 \cdot 5)$

2   A voltage supply has a maximum value of 100 V and a frequency of 60 Hz. Express the voltage V as a periodic function of the time t if at a time when $t = 0$ the voltage is (a) 0 V, (b) a maximum, (c) 50 V.

3   A weight on the end of a spring oscillates about a mean position. Its displacement y after a time t is given by $y = 3 \cdot 7 \sin (10 \cdot 4t + 0 \cdot 4)$. State the maximum displacement of the weight from its mean position and the time when this first occurs. State the frequency and the phase time of the motion. How long does it take for the spring to make one complete oscillation?

4   Show that $a \sin \theta + b \cos \theta$ may be expressed in the form $\sqrt{(a^2 + b^2)} \sin (\theta + \alpha)$ where $\tan \alpha = \dfrac{b}{a}$

Express $7 \sin 20\pi t + 24 \cos 20\pi t$ in the form $R \sin (20\pi t + \alpha)$ stating the amplitude, frequency and phase angle of the wave form.

5   Find the sum of the voltages $V_1 = 80 \sin 100\pi t$ and $V_2 = 60 \cos 100\pi t$ in the form $R \sin (100 \pi t + \alpha)$.

State the amplitude, period, frequency and phase angle of the resultant voltage.

Miscellaneous exercises 3

1   (a) If $z = 20 \left[\cos \left(\dfrac{\pi}{6} + \theta \right) + \cos \left(\dfrac{\pi}{6} - \theta \right) \right]$

find an expression for z in its simplest form.

(b) Express $6 \cos 5x \cos 9x$ as a sum.

(c) Show that $\dfrac{\sin A + \sin 3A + \sin 5A + \sin 7A}{\cos A + \cos 3A + \cos 5A + \cos 7A} = \tan 4A$.

2   (a) Express as products $\cos 48° + \cos 36°$ and $\sin 50° - \sin 30°$.

(b) Express as a sum or a difference $\sin 20° \cos 50°$ and $\sin 3x \sin 5x$.

(c) State the amplitudes and periods of the vibrations described by the following functions

$$y_1 = 2 \cdot 4 \sin (3 \cdot 5t + 1 \cdot 4) \text{ and } y_2 = 0 \cdot 9 \sin (3 \cdot 5t - 0 \cdot 8)$$

What is the phase difference between them?

3   (a) Show that $\cos A + \cos B = 2 \cos \frac{1}{2}(A + B) \cos \frac{1}{2}(A - B)$.

Hence show without using tables that $\cos 24° + \cos 96° + \cos 144° = 0$.

(b) Find all the positive values of $x < 180°$ for which $\cos 7x + \cos x = \cos 4x$.

U.E.I.

4   Expand $y = 6 \sin (2t - 30°) + 10 \sin (2t + 60°)$ and then express it in the form $y = R \sin (2t + \alpha)$.

What is the frequency and amplitude of this function?

5   (a) Express $\cos 2\theta - \cos 5\theta$ and $\sin 5\theta - \sin 2\theta$ each as a product and hence show that $\dfrac{\cos 2\theta - \cos 5\theta}{\sin 5\theta - \sin 2\theta} = \tan \dfrac{7\theta}{2}$.

(b) Express $16 \sin \theta - 21 \cos \theta$ in the form $R \sin (\theta - \phi)$ and hence find the value of θ to give a maximum value to $16 \sin \theta - 21 \cos \theta$.

U.L.C.I.

6   The displacement x m of a body from a fixed point about which it is oscillating is given by the expression $x = 4 \cdot 2 \sin 3t + 5 \cdot 4 \cos 3t$ where t is the time in seconds.

Express x in the form $R \sin (3t + \alpha)$ and state the amplitude and period of the oscillation.

Find the first time at which the body is at a distance 4 m from the fixed point.

7   A voltage supply has a maximum value of 60 V and a frequency of 40 Hz. Express the voltage V as a periodic function of the time t if at a time when $t = 0$ the voltage is (a) 0, (b) a maximum, (c) 30 V.

8   (a) Prove that

$$\frac{\cos 2\theta + \cos 6\theta}{\cos 4\theta} = \frac{\sin 3\theta - \sin \theta}{\sin \theta}$$

   (b) The angles being in radians, find the values of t between 0 and 2π for which

$$\cos \frac{t}{2} \cos \frac{3t}{2} + 0.44 = 0.$$

   E.M.E.U.

9   (a) If θ is an acute angle and $\tan \theta = \frac{3}{4}$ find without using tables the value of $3 \sin 2\theta + 4 \cos 2\theta$.
   (b) Without using tables show that

   $\sin 10° + \sin 20° + \sin 40° + \sin 50° = \sin 70° + \sin 80°$

10 (a) Express as a sum or difference

   i $\sin x \sin 2x \sin 3x$ ii $\sin (x + 30°) \cos (x - 30°)$

   (b) Prove $\dfrac{\sin 7\theta + \sin 5\theta}{\cos 5\theta - \cos 7\theta} = \cot \theta$.

   (c) State the amplitude, period, frequency and phase time of
   $x = 2.7 \sin (12t + 2.8)$

11 A spot of light oscillates along a straight line about a central point O. Its distance x from O is given by $x = 1.5 \sin 10\pi t$.
   What is the frequency? What are the first four times at which the body is at unit distance from O on either side of it?

12 (a) Prove $\dfrac{\sin \theta - 2 \sin^3 \theta}{2 \cos^3 \theta - \cos \theta} = \tan \theta$.

   (b) A and B are acute angles such that $\sin A + \sin B = 1.6$, $\cos A + \cos B = 1.2$.
   By squaring and adding these results find $\cos (A - B)$ and by division find $\tan \frac{1}{2}(A + B)$.
   Hence find values of A and B.

13 Prove that in any triangle

$$\sin \frac{A}{2} = \sqrt{\frac{(s-b)(s-c)}{bc}} \quad \text{and} \quad \cos \frac{A}{2} = \sqrt{\frac{s(s-a)}{bc}}$$

   where $s = \frac{1}{2}(a + b + c)$.

48   Compound angle formulae and periodic functions

Hence find an expression for the area of the triangle in terms of the sides.

Calculate the angles of a triangle of which the sides are 23·5, 38·2 and 54·7, and find its area.

D.T.C.

14 Find the angles and the area, in square metres, of the triangle of which the sides are, in metres, $a = 270, b = 372, c = 198$.

15 Prove that in any triangle $\tan \dfrac{B-C}{2} = \dfrac{b-c}{b+c} \cot \dfrac{A}{2}$.

By means of a range-finder the distances of two ships from a point of observation on shore were found to be 6310 and 4820 m, and their respective bearings were 72° 42′ E of S and 28° 20′ E of S. Calculate the distance apart of the ships and the bearing of the first as seen from the second.

D.T.C.

16 In an acute angled triangle $a = 312, c = 402$ and $A = 0.739$ rad. Find the other angles and side.

17 If $A = 0.812$ rad, $b = 76$ and $c = 22$, find the other angles and side and the area of the triangle.

4 Binomial, exponential and logarithmic series

4a Permutations and combinations

If we have a number of different objects which we can select and arrange in different orders, then the different arrangements are known as *permutations*. Thus the three letters a, b, c have the following different permutations when taken three at a time.

abc acb bac bca cab cba

When taken two at a time the possible permutations are

ab ba bc cb ac ca

If, however, we make a selection of the objects without regard to the order in the group then the different selections are known as *combinations*.

The possible combinations of the three letters a, b, c, when taken two at a time, total only three, namely ab, bc, ac. Notice that ab and ba are not different combinations but are different permutations.

The permutations of *n* different things, taken *r* at a time

The total number is denoted by $_nP_r$.

To find $_nP_r$ we think of the problem as that of having n different objects and r spaces or places to fill.

Since any one of the n objects can be put in the first space then this can be filled in n different ways. Having filled it there are $n-1$ different objects left and the second space can be filled in $n-1$ different ways. The third space can be filled in $n-2$ different ways with the $n-2$ objects left, and so on until the final place can be filled in $n-r+1$ different ways.

As each different way of filling the first place can be associated with each different way of filling the second place then the first two places can be filled in $n(n-1)$ different ways.

By similar reasoning the first three places can be filled in $n(n-1)(n-2)$ ways and all r places can be filled in $n(n-1)(n-2)\ldots(n-r+1)$ different ways.

Thus $_nP_r = n(n-1)(n-2)\ldots(n-r+1)$

$$= n(n-1)(n-2)\ldots(n-r+1) \times \frac{(n-r)(n-r-1)\ldots 2 \times 1}{(n-r)(n-r-1)\ldots 2 \times 1}$$

$$= \frac{n!}{(n-r)!}$$

where $n(n-1)(n-2)\ldots 3 \times 2 \times 1$ is denoted by $n!$ or $\lfloor n$ and is called *factorial n*.

Example 1 Find the number of permutations that can be formed from the six letters a, b, c, d, e, f, if taken

i   three at a time   ii four at a time   iii all together

i   The answer is the number of permutations of six objects taken three at a time,

$$_6P_3 = \frac{6!}{(6-3)!} = \frac{6 \times 5 \times 4 \times 3 \times 2 \times 1}{3 \times 2 \times 1} = 120$$

ii The number is

$$_6P_4 = \frac{6!}{(6-4)!} = \frac{6 \times 5 \times 4 \times 3 \times 2 \times 1}{2 \times 1} = 360$$

iii With all the letters the number is

$$6 \times 5 \times 4 \times 3 \times 2 \times 1 = 6! = 720$$

The formula $_6P_6$ gives $\dfrac{6!}{(6-6)!} = \dfrac{6!}{0!}$.

In order that the formula may be true we must take $0! = 1$.
We also obtain $0! = 1$ by putting $n = 1$ in $n! = n(n-1)!$

The combinations of n different things, taken r at a time

The total number is denoted by $_nC_r$.

If we take any one combination or group of r objects, these different objects can be re-arranged among themselves to give $_rP_r = r!$ different arrangements or permutations.

If all the $_nC_r$ groups are rearranged they would give a total of $_nC_r \times r!$ different permutations. This total is the total number of permutations from n objects taken r at a time.

Thus $_nC_r \times r! = {}_nP_r = \dfrac{n!}{(n-r)!}$

$$_nC_r = \frac{n!}{(n-r)!\, r!}$$

This number may also be written $\dbinom{n}{r}$.

Example 2 Find the number of ways of picking a team of four from a group of seven men.

The required number is the number of combinations of seven objects taken four at a time.

$$_7C_4 = \frac{7!}{(7-4)!\,4!} = \frac{7 \times 6 \times 5 \times 4 \times 3 \times 2 \times 1}{3 \times 2 \times 1 \times 4 \times 3 \times 2 \times 1}$$
$$= 35$$

Example 3 Evaluate $\binom{10}{7} + \binom{7}{2}$.

Answer $\dfrac{10!}{7!\,3!} + \dfrac{7!}{2!\,5!} = \dfrac{10 \times 9 \times \ldots \times 2 \times 1}{7 \times 6 \times \ldots \times 1 \times 3 \times 2 \times 1} + \dfrac{7 \times 6 \times \ldots \times 1}{2 \times 1 \times 5 \times 4 \times \ldots \times 1}$

$$= 120 + 21 = 141$$

Exercise 4a

1 Find $_5P_3$, $_7P_5$, $_7P_7$.

2 How many different permutations can be made from the letters of the word 'Monday' if taken

(a) three at a time?   (b) four at a time?

3 If there are eight horses in a race how many possible ways are there of nominating the first three?

4 Find (a) $_6C_3$, $_5C_2$, $_{10}C_4$

   (b) $\binom{7}{5}$, $\binom{6}{2}$, $\binom{6}{4}$

5 (a) Evaluate i $\binom{7}{3} + \binom{7}{2}$ ii $\binom{9}{6} + \binom{8}{3}$

   (b) Show that $_nC_r + {}_nC_{r-1} = {}_{n+1}C_r$

      when i $n = 6$, $r = 3$ and ii $n = 8$, $r = 4$

6 A guard of five has to be formed from eight men. How many different selections can be made?

7 How many different numbers can be formed from the digits 0, 1, 2, 3, 4 if each number contains all 5 digits?

8   A committee of 6 has to be chosen from 14 candidates. How many different committees can be chosen?

9   (a) Twelve football matches are selected as likely to result in drawn games. How many possible selections of 8 games can be made from the list of 12?
    (b) If three particular games are regarded as certain to result in drawn games and are always to be included in every selection of eight, how many different selections can now be made?

10 If the results of eight football matches have to be forecast as win, lose or draw for the home team, how many different forecasts are possible?

4b   The binomial theorem

The binomial theorem, already used in volume 1, is as follows:

$$(a+x)^n = a^n + na^{n-1}x + \frac{n(n-1)}{2!}a^{n-2}x^2 + \frac{n(n-1)(n-2)}{3!}a^{n-3}x^3 + \dots + x^n$$

where n is a positive integer.

To prove this result consider the expansion $(a+x)^n = (a+x)(a+x)(a+x)\dots$ to n factors.

Each term in the expansion must contain one letter from each bracket.

If a is chosen from each bracket we get the term a^n.

Choosing a from $n-1$ brackets and x from the remaining bracket gives $a^{n-1}x$. But the number of ways of doing this is the number of ways of selecting one bracket from n brackets.

i.e. $_nC_1 = \binom{n}{1}$. Therefore there are $\binom{n}{1}$ terms equal to $a^{n-1}x$.

If a is chosen from $n-2$ brackets and x from the remaining two we get $a^{n-2}x^2$. The number of ways of doing this is equal to the number of ways of choosing two brackets from n brackets, i.e. $\binom{n}{2}$.

The term containing x^r arises when a is selected from $n-r$ brackets and x from the remaining r brackets. The number of ways this occurs is equal to the number of ways of choosing r brackets from n brackets.

This gives the term $\binom{n}{r}a^{n-r}x^r$.

Finally, choosing x from all the n factors gives x^n.

Thus $(a+x)^n = a^n + \binom{n}{1}a^{n-1}x + \binom{n}{2}a^{n-2}x^2 + \dots + \binom{n}{r}a^{n-r}x^r + \dots + x^n$.

The general term or $(r+1)$th term

The first term $\qquad T_1 = a^n$

The second term $\quad T_2 = \dbinom{n}{1} a^{n-1}x$

The third term $\qquad T_3 = \dbinom{n}{2} a^{n-2}x^2$

The term containing x^r will be the $(r+1)$th term.

Thus $\qquad\qquad T_{r+1} = \dbinom{n}{r} a^{n-r}x^r$

Example 1 Expand $(2+x)^5$.

$$(2+x)^5 = 2^5 + \dbinom{5}{1} 2^4x + \dbinom{5}{2} 2^3x^2 + \dbinom{5}{3} 2^2x^3 + \dbinom{5}{4} 2x^4 + x^5$$
$$= 32 + 80x + 80x^2 + 40x^3 + 10x^4 + x^5$$

Example 2 Find the term containing x^6 in the expansion of $(2-x)^{10}$.

The required term is $\quad T_{r+1} = \dbinom{n}{r} a^{n-r}x^r \quad$ where $n = 10, r = 6$

i.e. $\qquad\qquad T_7 = \dbinom{10}{6} 2^{10-6}(-x)^6$

$$= \frac{10 \times 9 \times 8 \times 7 \times 6 \times 5}{1 \times 2 \times 3 \times 4 \times 5 \times 6} 16x^6$$
$$= 3360x^6$$

Exercise 4b

Expand:

1 $(a+x)^5$ $\qquad\qquad$ 2 $(2x+y)^4$ $\qquad\qquad$ 3 $(3x-5y)^3$

4 $\left(x+\dfrac{1}{x}\right)^6$ $\qquad\qquad$ 5 $(3x-y)^5$ $\qquad\qquad$ 6 $(1+x+x^2)^4$

Write down and simplify, leaving the answers in factorials,

7 the tenth term in the expansion of $(1-2x)^{12}$

8 the twelfth term in the expansion of $(2x-1)^{13}$

9 the fifth term in the expansion of $\left(2a-\dfrac{b}{3}\right)^8$

10   the seventh term in the expansion of $\left(\dfrac{4}{5}x-\dfrac{5}{2x}\right)^9$

11   the middle term in the expansion of $\left(x+\dfrac{a^2}{x}\right)^{12}$

12   the two middle terms in the expansion of $\left(2x-\dfrac{1}{x}\right)^9$.

4c   The binomial theorem for fractional or negative indices

The binomial theorem is true if the index n is a positive integer. When n is negative or fractional the series of terms goes on for ever and is called an *infinite* series. If, with such a series, the sum of the first n terms cannot exceed a certain value or limit S and continually approaches S as n increases, then the series is said to be *convergent* and S is the sum to infinity. Thus the infinite series $1+\frac{1}{2}+\frac{1}{4}+\frac{1}{8}+\ldots$ is convergent and has a sum to infinity of 2. (Note that it is a geometric progression with common ratio $\frac{1}{2}$.)

A series whose sum increases without limit as n increases is said to be *divergent*. In general for a series to be convergent the $(n+1)$th term must decrease indefinitely as n increases indefinitely and the sum of the remainder after n terms must become less than any assignable numerical quantity, however small, when n tends to infinity.

The binomial series becomes an infinite series for negative and fractional values of the index n and is not always convergent for all values of a and x. The left hand side is not necessarily equal to the right hand side for numerical work. For example expanding $\dfrac{1}{1+x}$ we get

$$\frac{1}{1+x} = (1+x)^{-1} = 1+(-1)\,x+\frac{-1.-2}{1.2}\,x^2+\frac{-1.-2.-3}{1.2.3}\,x^3+\ldots$$

$$= 1-x+x^2-x^3+\ldots \text{ where the terms continue indefinitely.}$$

If we put $x = 2$ we get

$$\tfrac{1}{3} = 1-2+4-8+16-\ldots$$

This is obviously an absurd result and arises because we have not made x smaller than unity.

For *fractional* and *negative indices* we express the binomial theorem in the form

$$(1+x)^n = 1+nx+\frac{n(n-1)}{2!}\,x^2+\frac{n(n-1)(n-2)}{3!}\,x^3+\ldots$$

which is true provided that x is numerically less than unity. This is written

$x < 1$ numerically, or $-1 < x < 1$ or $|x| < 1$, where $|x|$ is read as 'the numerical value of x'.

Example 1   Expand each of the following in ascending powers of x as far as the term in x^3 giving the limits of x for which each expansion is valid.

i $\dfrac{1}{(1+2x)^3}$ ii $\dfrac{1}{(8-x)^2}$ iii $\sqrt{(4+x)}$

i $\dfrac{1}{(1+2x)^3} = (1+2x)^{-3} = 1+(-3)\,2x+\dfrac{-3 \times -4}{1 \times 2}(2x)^2$

$$+\dfrac{-3 \times -4 \times -5}{1 \times 2 \times 3}(2x)^3 + \ldots$$

$$= 1-6x+24x^2-80x^3 + \ldots$$

This is true provided $2x < 1$ numerically
i.e. $x < \frac{1}{2}$ numerically.

ii $\dfrac{1}{(8-x)^2} = \dfrac{1}{\left[8\left(1-\dfrac{x}{8}\right)\right]^2} = \dfrac{1}{8^2\left(1-\dfrac{x}{8}\right)^2}$

$$= \dfrac{1}{64}\left(1-\dfrac{x}{8}\right)^{-2} = \dfrac{1}{64}\left[1+\dfrac{-2}{1}\left(\dfrac{-x}{8}\right)+\dfrac{-2 \times -3}{1 \times 2}\left(\dfrac{-x}{8}\right)^2 +\right.$$

$$\left.+\dfrac{-2 \times -3 \times -4}{1 \times 2 \times 3}\left(\dfrac{-x}{8}\right)^3 + \ldots\right]$$

$$= \dfrac{1}{64}\left[1+\dfrac{x}{4}+\dfrac{3x^2}{64}+\dfrac{x^3}{128}\ldots\right]$$

provided $\dfrac{x}{8} < 1$ numerically

i.e. $x < 8$ numerically.

iii $\sqrt{(4+x)} = \sqrt{\left[4\left(1+\dfrac{x}{4}\right)\right]} = 2\left(1+\dfrac{x}{4}\right)^{\frac{1}{2}}$

$$= 2\left[1+\dfrac{1}{2}\times\dfrac{x}{4}+\dfrac{\frac{1}{2}\times-\frac{1}{2}}{1 \times 2}\left(\dfrac{x}{4}\right)^2+\dfrac{\frac{1}{2}\times-\frac{1}{2}\times-\frac{3}{2}}{1 \times 2 \times 3}\left(\dfrac{x}{4}\right)^3 + \ldots\right]$$

$$= 2+\dfrac{x}{4}-\dfrac{x^2}{64}+\dfrac{x^3}{512} - \ldots$$

provided $\dfrac{x}{4} < 1$ numerically

i.e. $x < 4$ numerically

Example 2 Express $\dfrac{3x-1}{3+2x-x^2}$ as the sum of two partial fractions. Hence expand the expression as a series in ascending powers of x as far as the term in x^3, giving the values of x for which the expansion is valid.

Let $\dfrac{3x-1}{3+2x-x^2} \equiv \dfrac{A}{3-x} + \dfrac{B}{1+x}$

$$\equiv \frac{A(1+x)+B(3-x)}{(3-x)(1+x)}$$

Equating numerators and putting $x = 3$ we get

$8 = 4A$
$A = 2$

Putting $x = -1$ gives $-4 = 4B$
$$B = -1$$

Thus $\dfrac{3x-1}{3+2x-x^2} = \dfrac{2}{3-x} - \dfrac{1}{1+x}$

Taking the first fraction:

$$\frac{2}{3-x} = \frac{2}{3\left(1-\dfrac{x}{3}\right)}$$

$$= \frac{2}{3}\left(1-\frac{x}{3}\right)^{-1}$$

$$= \frac{2}{3}\left[1+(-1)\left(-\frac{x}{3}\right)+\frac{-1\times-2}{1\times2}\left(-\frac{x}{3}\right)^2+\frac{-1\times-2\times-3}{1\times2\times3}\left(-\frac{x}{3}\right)^3+\ldots\right]$$

$$= \frac{2}{3}\left(1+\frac{x}{3}+\frac{x^2}{9}+\frac{x^3}{27}+\ldots\right)$$

provided $x < 3$ numerically.

Expanding the second fraction:

$$\frac{1}{1+x} = (1+x)^{-1} = 1+(-1)x+\frac{-1\times-2}{1\times2}x^2+\frac{-1\times-2\times-3}{1\times2\times3}x^3+\ldots$$

$$= 1-x+x^2-x^3+\ldots$$

provided $x < 1$ numerically.

Thus $\dfrac{3x-1}{3+2x-x^2} = -\dfrac{1}{3}+\dfrac{11}{9}x-\dfrac{25}{27}x^2+\dfrac{83}{81}x^3+\ldots$

provided $x < 1$ numerically.

57 **The binomial theorem for negative or fractional indices**

Exercise 4c

Expand the following in ascending powers of x as far as the term in x^3 giving the limits of x for which the expansions are true.

1 $\dfrac{1}{1-x}$ 2 $\dfrac{1}{(1-x)^2}$ 3 $\dfrac{1}{(1+x)^2}$

4 $\sqrt{(1+x)}$ 5 $\dfrac{1}{\sqrt{(1-2x)}}$ 6 $\dfrac{1}{\sqrt{(4-x)}}$

7 $\dfrac{1}{(3+x)^3}$ 8 $(9-2x)^{\frac{3}{2}}$ 9 $\dfrac{1}{\sqrt{(4+x)}}$

10 $\dfrac{1}{3+2x}$

Express the following in partial fractions. Hence expand each expression as a series in ascending powers of x as far as the term in x^3, giving the range of values of x for which the expansion is valid.

11 $\dfrac{3-x}{(1-x)(1+x)}$ 12 $\dfrac{1-7x}{(2+x)(1-2x)}$

13 $\dfrac{2-5x-4x^2}{(1+x)(1-2x)^2}$ 14 $\dfrac{3x-5}{(x-2)(x-1)^2}$

4d Approximations

If x is small we can find the value of $(1+x)^n$ to any degree of accuracy by the binomial expansion.

Since $(1+x)^n = 1+nx+\dfrac{n(n-1)}{2!}x^2+\dots$, a rough approximation is obtained by rejecting all powers of x above the first. Thus $(1+x)^n = 1+nx$ approximately.

If a closer approximation is required we can retain the term containing x^2 and reject higher powers – and so on according to the degree of accuracy required.

Example 1 Find approximate values for:

i $(1\cdot02)^3$ ii $\sqrt[3]{0\cdot998}$ iii $\dfrac{1}{1\cdot005}$

$(1+x)^n = 1+nx+\dfrac{n(n-1)}{2!}x^2$ approximately

i $(1·02)^3 = (1+0·02)^3 \simeq 1+3 \times 0·02 + 3 \times 0·0004$

$$\simeq 1·0612$$
$$= 1·06 \text{ with approximate error } 0·001$$

ii $\sqrt[3]{0·998} = (1-0·002)^{\frac{1}{3}} \simeq 1 - \frac{1}{3} \times 0·002 + \frac{\frac{1}{3} \times -\frac{2}{3}}{1 \times 2}(0·002)^2$

$$\simeq 1 - 0·000\,666\,7 - 0·000\,000\,4$$
$$= 0·999\,33 \text{ with approximate error } 0·000\,004$$

iii $\dfrac{1}{1·005} = (1+0·005)^{-1} \simeq 1 + (-1)\,0·005 + \dfrac{-1 \times -2}{1 \times 2}(0·005)^2$

$$\simeq 1 - 0·005 + 0·000\,025$$
$$= 0·9950 \text{ with approximate error } 0·000\,03$$

Example 2   Find $\sqrt[4]{623}$ correct to six decimal places.

Since $5^4 = 625$ we write

$$\sqrt[4]{623} = (625-2)^{\frac{1}{4}} = 625^{\frac{1}{4}}\left(1 - \frac{2}{625}\right)^{\frac{1}{4}}$$

$$= 5\left[1 + \frac{1}{4}\left(\frac{-2}{625}\right) + \frac{\frac{1}{4} \times -\frac{3}{4}}{1 \times 2}\left(\frac{-2}{625}\right)^2 + \frac{\frac{1}{4} \times -\frac{3}{4} \times -\frac{7}{4}}{1 \times 2 \times 3}\left(\frac{-2}{625}\right)^3 + \ldots\right]$$

$$= 5\left[1 - \frac{0·0032}{4} - \frac{3}{32}(0·0032)^2 - \frac{7}{128}(0·0032)^3 \ldots\right]$$

$$= 5\,[1 - 0·0008 - 0·000\,000\,96 - 0·000\,000\,002 \ldots]$$

$$\simeq 5\,[1 - 0·000\,800\,96]$$

$= 4·995\,995$ correct to six decimal places as including the next term in the expansion would not alter any of these figures.

Note that if we used the approximation $(1+x)^n \simeq 1+nx$ we would get $\sqrt[4]{623} = 4·9960$ which is correct to four decimal places.

Example 3   If x be small compared with unity find an approximate value for:

i $(1+x)^{\frac{2}{3}}(1-x)^{\frac{1}{2}}$ ii $\dfrac{(1+x)^{\frac{2}{3}}\sqrt{(1-2x)}}{\sqrt[3]{(1+x)}}$

When α and β are both small fractions.

$$(1+\alpha)^m (1+\beta)^n = (1+m\alpha)(1+n\beta) \text{ approximately}$$
$$= 1 + m\alpha + n\beta \text{ approximately.}$$

Thus i $(1+x)^{\frac{2}{3}}(1-x)^{\frac{1}{2}} \simeq 1 + \frac{2}{3}x - \frac{1}{2}x$

$$\simeq 1 + \frac{x}{6}$$

and ii $(1+x)^{\frac{2}{3}}(1-2x)^{\frac{1}{2}}(1+x)^{-\frac{1}{3}} \simeq 1 + \frac{2}{3}x + \frac{1}{2}(-2x) - \frac{1}{3}x$

$$\simeq 1 - \frac{2}{3}x$$

Example 4 The time of swing of a simple pendulum is given by the formula $T = 2\pi\sqrt{\dfrac{l}{g}}$ where l is its length and g is the acceleration of gravity. If in a timing experiment l is measured $1\frac{1}{2}\%$ too large and T is measured 2% too small, find the approximate percentage error this would cause in calculating g.

Rearranging the formula we get $g = \dfrac{4\pi^2 l}{T^2}$

If g, l and T are the correct values then the calculated value of g,

$$g' = \frac{4\pi^2 l\left(1+\dfrac{1\frac{1}{2}}{100}\right)}{\left[T\left(1-\dfrac{2}{100}\right)\right]^2}$$

$$= \frac{4\pi^2 l}{T^2}\left(1+\frac{3}{200}\right)\left(1-\frac{2}{100}\right)^{-2}$$

$$\simeq \frac{4\pi^2 l}{T^2}\left(1+\frac{3}{200}+\frac{-2}{100}\times -2\right)$$

$$\simeq g\left(1+\frac{11}{200}\right)$$

Thus the calculated value is $\dfrac{11}{200}$ or $5\frac{1}{2}\%$ too large.

Exercise 4d
When x is very small show that

1   $\dfrac{1}{(1-x)^2\sqrt{(1-x)}} \simeq 1+\dfrac{5}{2}x$

2   $\dfrac{1}{(1-3x)^2(1-2x)^3} \simeq 1+12x$

3   $\dfrac{1+x}{(1-4x)^5} \simeq 1+21x$

4   $\dfrac{\sqrt{(1+x)}}{\sqrt[3]{(1+x)}} \simeq 1+\dfrac{x}{6}$

5   $\dfrac{(4+x)^{\frac{1}{2}}}{(3-x)^2} \simeq \dfrac{2}{9}\left(1+\dfrac{19}{24}x\right)$

Find to four places of decimals:

6   $\sqrt{98}$ 7 $\sqrt[3]{1003}$ 8 $\sqrt[3]{998}$ 9 $\sqrt[3]{510}$

10   Show that if x^3 and higher powers of x can be neglected

$$\frac{(1+x)^4}{\sqrt{(1-x)}} = 1+\frac{9}{2}x+\frac{67}{8}x^2$$

Use this expansion to evaluate $\dfrac{(1\cdot02)^4}{\sqrt{0\cdot98}}$

11 An error of 0.5% is made in measuring the diameter of a sphere. Show that the approximate percentage error in the volume is 1.5%.

12 In the formula $h = \dfrac{klv^2}{d}$ find the percentage error in h due to an error of $+3\%$ in l, $-1\frac{1}{2}\%$ error in v and $+1\%$ error in d.

13 Show that $30 = 3^3\,(1+\frac{1}{9})$ and hence evaluate $\sqrt[3]{30}$ correct to three places of decimals.

14 A formula for power wasted in air friction is $P = cd^{5.5}n^{3.5}$. Find the percentage change in P due to a 2% increase in d and a 1% fall in n.

15 Show that if x is small compared to a

(a) $\sqrt{(a+x)} \simeq \sqrt{a}\left(1 + \dfrac{x}{2a}\right)$ (b) $\dfrac{1}{a-x} \simeq \dfrac{1}{a} + \dfrac{x}{a^2}$

Using the result in (b) find an approximate value for $\dfrac{1}{4.95}$.

16 Express $\sqrt{17}$ in the form $4\sqrt{(1+a)}$. Hence find the value of $\sqrt{17}$ correct to four significant figures.

17 The shear stress q in a shaft of diameter D under a torque T is given by $q = \dfrac{16T}{\pi D^3}$. Find the approximate percentage error in calculating q if T is measured 2% too small and D $1\frac{1}{2}\%$ too large.

18 Expand $(1-2x)^{\frac{1}{2}}$ and $(1+x)^{-3}$ each as far as the term in x^2. Hence expand $\dfrac{\sqrt{(1-2x)}}{(1+x)^3}$ as far as the term in x^2 stating the value of x for which the expansion is valid.

19 The deflexion at the centre of a rod is given by $y = \dfrac{KWl^3}{d^4}$. Find the approximate percentage change in the deflexion if W increases by 3%, l decreases by 2% and d increases by 1%.

20 If x is very small such that x^2 and higher powers may be neglected, find approximate values for

(a) $\dfrac{\sqrt[3]{(8-2x)}\,\sqrt{(x+4)}}{(1+x)^{\frac{2}{3}}}$ (b) $\dfrac{\sqrt{(9-x)}(1+2x)^{\frac{3}{2}}}{\sqrt[4]{(81+3x)}}$

61 Approximations

4e   The exponential series

The binomial series can be used to derive the expansion for an important function denoted by e^x and known as the exponential function. It may be defined as being $\lim_{n \to \infty} \left(1 + \dfrac{x}{n}\right)^n$

Expanding by the binomial theorem

$$\left(1 + \frac{x}{n}\right)^n = 1 + n\left(\frac{x}{n}\right) + \frac{n(n-1)}{2!}\left(\frac{x}{n}\right)^2 + \frac{n(n-1)(n-2)}{3!}\left(\frac{x}{n}\right)^3 + \dots \text{ to infinity.}$$

$$= 1 + x + \frac{\left(1 - \dfrac{1}{n}\right)}{2!}x^2 + \frac{\left(1 - \dfrac{1}{n}\right)\left(1 - \dfrac{2}{n}\right)}{3!}x^3 + \dots$$

This is true for values of n however great and therefore true when n is infinite. If we now let n tend to infinity and note that $1 - \dfrac{1}{n}, 1 - \dfrac{2}{n}, \dots$ all tend to 1 we get $e^x = 1 + x + \dfrac{x^2}{2!} + \dfrac{x^3}{3!} + \dots + \dfrac{x^r}{r!} + \dots$ to infinity.

[The argument involving the limits of the terms and the sum of the limits requires more careful examination than is possible here.

An alternative method of obtaining the exponential function e^x without this difficulty is given in the next chapter.]

The series for e^x is true for all values of x positive or negative.

Putting $x = 1$

$$e = 1 + 1 + \frac{1}{2!} + \frac{1}{3!} + \frac{1}{4!} + \dots$$

This series is convergent since each term after the first is less than the corresponding term of the series.

$$1 + 1 + \frac{1}{2} + \frac{1}{2^2} + \frac{1}{2^3} + \dots = 3$$

which is, after the first term, a geometric progression.

To five decimal places $e = 2 \cdot 718\,28$. Like π, e is incommensurable, having a non-terminating and non-recurring decimal part. It is the base of natural (or Napierian) logarithms.

Replacing x by $-x$ in the expansion for e^x

$$e^{-x} = 1 - x + \frac{x^2}{2!} - \frac{x^3}{3!} + \frac{x^4}{4!} \dots$$

Note $e = 1 + 1 + \dfrac{1}{2!} + \dfrac{1}{3!} + \dfrac{1}{4!} + \dots$

$$e^{-1} = 1 - 1 + \frac{1}{2!} - \frac{1}{3!} + \frac{1}{4!} - \cdots$$

$$e^{2} = 1 + 2 + \frac{2^2}{2!} + \frac{2^3}{3!} + \frac{2^4}{4!} + \cdots$$

$$e^{-2} = 1 - 2 + \frac{2^2}{2!} - \frac{2^3}{3!} + \frac{2^4}{4!} - \cdots$$

Example 1 Show that $\dfrac{e^2 + 1}{2e} = 1 + \dfrac{1}{2!} + \dfrac{1}{4!} + \dfrac{1}{6!} + \cdots$

$$\begin{aligned}
\frac{e^2 + 1}{2e} &= \frac{1}{2}\left(e + \frac{1}{e}\right) = \frac{1}{2}(e + e^{-1}) \\
&= \frac{1}{2}\left(1 + 1 + \frac{1}{2!} + \frac{1}{3!} + \frac{1}{4!} + \cdots + 1 - 1 + \frac{1}{2!} - \frac{1}{3!} + \frac{1}{4!} - \cdots\right) \\
&= \frac{1}{2}\left(2 + \frac{2}{2!} + \frac{2}{4!} + \frac{2}{6!} + \cdots\right) \\
&= 1 + \frac{1}{2!} + \frac{1}{4!} + \frac{1}{6!} + \cdots
\end{aligned}$$

If we plot the graph of $y = e^x$ for positive and negative values of x we get the curve shown.

x	0	1	2	3	-1	-2	-3
e^x	1	2·72	7·39	20·0	0·37	0·14	0·05

Figure 5

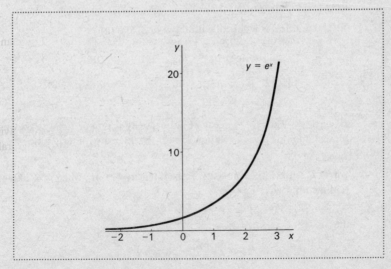

Students should be able to use the tables of e^x and of logarithms to base e (natural or hyperbolic logarithms) which are found in most sets of mathematical tables.

Example 2 From the tables find
i $\log_e 4 \cdot 351$ ii $\log_e 435 \cdot 1$ iii $\log_e 0 \cdot 004 \, 351$.

i From the table obtain $\log_e 4 \cdot 351 = 1 \cdot 4704$.
ii $\qquad 435 \cdot 1 = 4 \cdot 351 \times 10^2$
$\quad \log_e 435 \cdot 1 = \log_e 4 \cdot 351 + 2 \log_e 10$
$\qquad\qquad\qquad = 1 \cdot 4704 + 2 \times 2 \cdot 3026 = 6 \cdot 0756$
iii $\qquad 0 \cdot 004 \, 351 = 4 \cdot 351 \times 10^{-3}$
$\quad \log_e 0 \cdot 004 \, 351 = \log_e 4 \cdot 351 + \log_e 10^{-3}$
$\qquad\qquad\qquad = 1 \cdot 4704 + \bar{7} \cdot 0922 \text{ (using the table for } \log_e 10^{-n})$
$\qquad\qquad\qquad = \bar{6} \cdot 5626 \text{ or } -5 \cdot 4374$

Example 3 Find the numbers whose natural logarithms are
i $\bar{7} \cdot 3641$ ii $\bar{6} \cdot 4567$
iii If $\log_e x = -6 \cdot 6324$ find x.

i $\qquad\qquad \bar{7} \cdot 3641 = 6 \cdot 9078 + 0 \cdot 4563$
$\qquad\qquad\qquad = \log_e 10^3 + \log_e 1 \cdot 578$
$\qquad\qquad\qquad = \log_e 1578$
Required number $= 1578$

ii $\qquad\qquad \bar{6} \cdot 4567 = \bar{7} \cdot 0922 + 1 \cdot 3645$
$\qquad\qquad\qquad = \log_e 10^{-3} + \log_e 3 \cdot 914$
$\qquad\qquad\qquad = \log_e 0 \cdot 003 \, 914$
Required number $= 0 \cdot 003 \, 914$

iii The problem is identical with those in (i) and (ii)
$\qquad\qquad -6 \cdot 6324 = -6 \cdot 9078 + 0 \cdot 2754$
$\qquad\qquad\qquad = \log_e 10^{-3} + \log_e 1 \cdot 317$
$\qquad\qquad\qquad = \log_e 0 \cdot 001 \, 317$
$\qquad\qquad \log_e x = \log_e 0 \cdot 001 \, 317$
$\qquad\qquad\qquad x = 0 \cdot 001 \, 317$

Example 4 The rate of decay of radioactivity is proportional to the amount present and this gives a relationship for the intensity I after a time t as
$$I = I_0 \, e^{-kt}$$
where I_0 is the initial intensity. If the intensity $I = \frac{1}{2} I_0$ when $t = 2000$ years, find the time when $I = \frac{1}{10} I_0$.

When $\quad t = 2000 \quad I = \frac{1}{2} I_0$
thus $\qquad\qquad \frac{1}{2} I_0 = I_0 \, e^{-k \times 2000}$
$$\frac{1}{2} = e^{-2000k} = \frac{1}{e^{2000k}}$$

or $\qquad e^{2000k} = 2$

Thus $\qquad 2000k = \log_e 2 = 0{\cdot}6931$

Let T be the time for the intensity to fall to $\frac{1}{10}I_0$.

Then $\qquad \frac{1}{10}I_0 = I_0\, e^{-kT}$

$$\frac{1}{10} = e^{-kT} = \frac{1}{e^{kT}}$$

$$kT = \log_e 10 = 2{\cdot}3026$$

$$T = \frac{2{\cdot}3026}{k} = \frac{2{\cdot}3026 \times 2000}{0{\cdot}6931}$$

$$= 6645 \text{ years}$$

4f The logarithmic series

If in the expansion $e^x = 1 + x + \dfrac{x^2}{2!} + \dfrac{x^3}{3!} + \ldots$ we write $e^x = a^y$ then taking logs

to base e

$\log_e(e^x) = \log_e(a^y)$

i.e. $\quad x = y \log_e a$

and the expansion may be written

$$a^y = 1 + y \log_e a + \frac{(y \log_e a)^2}{2!} + \frac{(y \log_e a)^3}{3!} + \ldots$$

Now writing $1 + z$ for a we get

$$(1+z)^y = 1 + y \log_e(1+z) + \frac{[y \log_e(1+z)]^2}{2!} + \frac{[y \log_e(1+z)]^3}{3!} + \ldots$$

Provided z be numerically less than unity $(1+z)^y$ can be expanded by the binomial theorem giving

$$1 + yz + \frac{y(y-1)}{2!}z^2 + \frac{y(y-1)(y-2)}{3!}z^3 + \ldots$$

$$= 1 + y \log_e(1+z) + \frac{[y \log_e(1+z)]^2}{2!} + \ldots$$

Equating the coefficients of y on the two sides of the equation we get

$$\log_e(1+z) = z - \frac{z^2}{2} + \frac{z^3}{3} - \ldots$$

This series is known as the *logarithmic series* and is true provided z is numerically less than unity. Thus

$$\log_e(1+x) = x - \frac{x^2}{2} + \frac{x^3}{3} - \frac{x^4}{4} + \ldots \text{ to infinity} \quad \text{if } |x| < 1.$$

Similarly writing $-x$ for x in the above

$$\log_e(1-x) = -x - \frac{x^2}{2} - \frac{x^3}{3} - \frac{x^4}{4} - \ldots \text{ to infinity} \quad \text{if } |x| < 1.$$

Example 1 Expand $\log_e(1+7x+12x^2)$ in ascending powers of x and state the limits within which the expansion is valid.

$$\log_e(1+7x+12x^2) = \log_e(1+3x)(1+4x)$$
$$\log_e(1+3x) + \log_e(1+4x)$$

$$= 3x - \frac{(3x)^2}{2} + \frac{(3x)^3}{3} - \ldots (\text{if } |3x| < 1) \ldots$$

$$\ldots + 4x - \frac{(4x)^2}{2} + \frac{(4x)^3}{3} - \ldots (\text{if } |4x| < 1)$$

$$= 7x - \frac{25}{2}x^2 + \frac{91}{3}x^3 - \ldots \text{ provided } x < \tfrac{1}{4} \text{ numerically.}$$

In calculating logarithms of numbers the ordinary expansion is not suitable as we must evaluate too many terms, making the work tedious. A better series which can be used is obtained as follows:

$$\log_e \frac{1+x}{1-x} = \log_e(1+x) - \log_e(1-x)$$

$$= \left(x - \frac{x^2}{2} + \frac{x^3}{3} - \ldots \right) - \left(-x - \frac{x^2}{2} - \frac{x^3}{3} - \ldots \right)$$

$$= 2\left(x + \frac{x^3}{3} + \frac{x^5}{5} + \ldots \right)$$

If we now write $\dfrac{1+x}{1-x} = \dfrac{n+1}{n}$

$$n + nx = n + 1 - nx - x$$

and $$x = \frac{1}{2n+1}$$

Hence $$\log_e \frac{n+1}{n} = 2\left[\frac{1}{2n+1} + \frac{1}{3(2n+1)^3} + \frac{1}{5(2n+1)^5} + \ldots \right]$$

where the expansion is true provided $n > 0$.

This series enables us to calculate $\log_e 2$ by putting $n = 1$. Then by putting $n = 2$ we can get $\log_e 3$ from $\log_e 2$, and so on.

These can be converted into common logarithms by multiplying by

$$\log_{10} e = \frac{1}{\log_e 10} = 0{\cdot}434\,29$$

The general shape of the graph of $y = \log_e x$ is indicated in fig. 6.

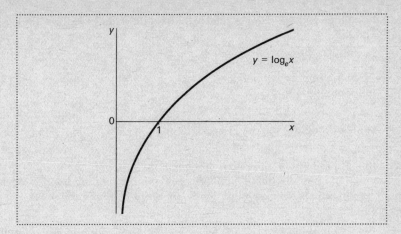

Exercise 4f

Expand each of the following (1–10) in ascending powers of x as far as the fourth term giving the limits of x for which the expansion is valid.

1 $\sqrt{e^x}$

2 $(1+x^2)\,e^x$

3 $\dfrac{(1-x)^2}{e^x}$

4 $\dfrac{e^{2x}}{\sqrt{(1-x)}}$

5 $\log_e (1+2x)$

6 $\log_e (1+x)^2$

7 $\log_e 2(1+x)$

8 $\log_e (1+5x+6x^2)$

9 $\log_e \dfrac{1-2x}{1-3x}$

10 $(1+3x+x^2)\,e^x$

11 (a) From the tables find i $\log_e 756{\cdot}3$ ii $\log_e 0{\cdot}005\,165$
    (b) Evaluate i $e^{1{\cdot}337}$ ii $e^{7{\cdot}629}$ iii $e^{-6{\cdot}321}$
    (c) If $\log_e x = 4{\cdot}3848$ find x.

12 Quote the series for e^x and expand to four terms:

    (a) $\dfrac{1}{2}\left(e-\dfrac{1}{e}\right)$ (b) $\sqrt{e}+\dfrac{1}{\sqrt{e}}$

13 Calculate correct to three decimal places:

    (a) $\dfrac{e^{0{\cdot}1}+e^{-0{\cdot}1}}{2}$ and   (b) $\dfrac{e^{0{\cdot}1}-e^{-0{\cdot}1}}{2}$

67 **The logarithmic series**

14   Given $\log_e 2 = 0.693\,147$ find $\log_e 3$ from the series for $\log_e (n+1) - \log_e n$.

15   Solve $3e^{2x} - 7e^x + 2 = 0$.

16   Solve $6e^x + 2e^{-x} = 7$.

17   Solve $\log_e x + \log_e (x+3) = 2.3026$.

18   Expand $e^{3x} + e^{-3x}$ as far as the term in x^4.
      If x is small so that x^4 and higher powers may be neglected, use the expansion to find approximate values for the roots of $e^{3x} + e^{-3x} - 2.09 = 0$.

19   If x is so small that x^4 and higher powers of x may be neglected, show that

$$(x^2 + 2x + 2)\, e^{-x} = 2 - \frac{x^3}{3} \text{ approximately.}$$

20   Give the first three terms of the series for e^{2x} and $\dfrac{1}{(1-x)^3}$ in ascending powers of x.

      Hence evaluate $\dfrac{e^{2x}}{(1-x)^3}$ to four significant figures when $x = 0.01$.

Miscellaneous exercises 4

1   (a) Expand $\dfrac{1}{\sqrt{(1-2x)}}$ as a series in ascending powers of x to four terms and state for what range of values of x the expansion is true.

      (b) If x is small so that terms containing x^2 and higher powers may be neglected find an approximate value for $\dfrac{\sqrt{(1+4x)}(1-\frac{1}{2}x)^3}{(1+x)^2}$.

2   (a) Write down the first four terms of the expansion of $(8+3x)^{\frac{1}{3}}$ in ascending powers of x and state the range of values of x for which the expansion is valid.

      (b) If x is small show that $\dfrac{\sqrt{(9-x)}(1+2x)^{\frac{1}{3}}}{\sqrt[4]{(81+3x)}} \simeq 1 + \dfrac{65}{108}\, x$.

      Use this result to find the approximate value of the expression when $x = 0.108$.

3   (a) If h is small compared to x find approximate values for

$$\text{i}\ \frac{(x+h)^5-x^5}{h} \qquad \text{ii}\ \frac{(x+h)^{\frac{1}{4}}-x^{\frac{1}{4}}}{h} \qquad \text{iii}\ \frac{\frac{1}{(x+h)^2}-\frac{1}{x^2}}{h}$$

(b) Evaluate the $\displaystyle\lim_{x\to 1}\frac{\log_e x}{x-1}$ by putting $x=1+h$.

4   Express $\dfrac{3+5x}{(1+x)(1+2x)}$ as the sum of two partial fractions. Use the binomial theorem to expand the expression as a series in ascending powers of x as far as the term in x^3. For what values of x is the expansion valid?

5   (a) Find the first four terms in the binomial expansion of:

i $(1+x)^{\frac{1}{2}}$ and   ii $(1-x)^{-\frac{1}{2}}$

and obtain the expansion for $\sqrt{\dfrac{1+x}{1-x}}$ as far as the term in x^2.

(b) The time of one swing (T) of a pendulum of length L is given by $T=k\sqrt{L}$ where k is a constant. Use the binomial theorem to find the percentage change in the value of T if L is increased by 1%. Express your answer correct to two decimal places.

N.C.T.E.C.

6   (a) Obtain the binomial expansions in ascending powers of x as far as the terms containing x^3 of $(1-x)^4$ and of $(2+x)^8$.
    Evaluate A, B, C, D if

$$(1-x)^4(2+x)^8 \equiv A+Bx+Cx^2+Dx^3+ \ldots$$

(b) Prove that, for sufficiently small values of x,

$$\sqrt{(100-4x^2)} = 10-0{\cdot}2x^2-0{\cdot}002x^4- \ldots$$

and deduce the value of $\sqrt{96}$ correct to three decimal places.

E.M.E.U.

7   (a) Evaluate i $_9C_4+{_7}C_3$ $\qquad$ ii $\dbinom{8}{3}+\dbinom{8}{2}$

(b) A team of six has to be chosen from ten players. How many different teams can be chosen?

69   Miscellaneous exercises 4

8   (a) For the binomial expansion of $(1+x)^n$, where n is not a positive integer, state the range of values of x for which the expansion is valid. If n is a positive integer, how many terms are there in the expansion?
     Find the value of the term independent of x in the expansion of $\left(2x+\dfrac{1}{5x}\right)^6$.

    (b) Show that $\left(1+\dfrac{7x}{2}\right)^{\frac{2}{5}} = 1+\dfrac{7}{5}x-\dfrac{147}{100}x^2+\ldots$ and hence evaluate $(1\cdot035)^{\frac{2}{5}}$ correct to four decimal places.

    (c) The formula $H = \dfrac{M}{2h}\left[\dfrac{1}{(x-h)^2}-\dfrac{1}{(x+h)^2}\right]$ expresses the field strength due to a magnet of length $2h$ and moment M at a point distant x from its centre. If h is small compared with x, show that

$$H \simeq \frac{2M}{x^3}\left(1+\frac{2h^2}{x^2}\right)$$

U.E.I.

9   Solve the following for x.

    (a) $e^{-x^2} = 0\cdot5$
    (b) $e^{-\sin x} = 0\cdot447$
    (c) $\log_e (x^2+0\cdot5) = 0\cdot27$
    (d) $\log_e x - \log_e (x-1) = 1$

10 (a) In the formula $I = 100\,(1-e^{-Rt/L})$ calculate I when $R = 100$, $L = 2$ and $t = 0\cdot05$.
    (b) The current i A in a circuit at a time t s is given by $i = i_0\,e^{-Rt/L}$. If $R = 12$ and $L = 0\cdot25$ find the time taken for the current to fall from 15 A to 5 A.

11 (a) Show $3\sqrt{13} = 10(1+17/100)^{\frac{1}{2}}$.
     Hence, evaluate $\sqrt{13}$ to three decimal places.
    (b) The effective range of a radio antenna is given by $R = kp^{\frac{1}{4}}/f$ where k is a constant, p is the antenna power and f is the frequency. If this frequency is increased by 4% and the antenna power reduced by 3% find, correct to one decimal place, the percentage change in the effective range.

U.E.I.

12 (a) Write down the first three terms in the expansions of $(1-x^2)^{\frac{1}{2}}$ and $(1+x^2)^{-\frac{1}{2}}$ stating for what values of x these are valid.
     Use your results to evaluate $\sqrt{\dfrac{0\cdot99}{1\cdot01}}$ correct to five decimal places.

(b) If $P = \dfrac{Wb^3}{a^4 c^{\frac{3}{2}}}$ find, using first order binomial approximation, the percentage change in the value of P when b increases by 0.3%, a increases by 0.8% and c increases by 0.2%.

N.C.T.E.C.

13 Write down the series for $\log_e (1+x)$ and deduce that

$$\log_e \frac{1+x}{1-x} = 2\left[x + \frac{x^3}{3} + \frac{x^5}{5} + \cdots \right]$$

By putting $x = \frac{1}{3}$ in the expansion, find the value of $\log_e 2$ correct to four places of decimals.

14 State without proof the series for e^x and $\log_e (1+x)$.

If $y = x - \dfrac{x^2}{2} + \dfrac{x^3}{3} - \dfrac{x^4}{4} + \cdots$

show that

$$x = y + \frac{y^2}{2!} + \frac{y^3}{3!} + \frac{y^4}{4!} + \cdots$$

State the limits between which x should lie for the above expressions to be true, and the corresponding limits of y.

D.T.C.

15 (a) Write down the coefficient of x^r in the expansion of $(1+x)^n$, and hence show that the $(r+1)$th term of the expansion of $(1-x)^{-2}$ is $(r+1)\,x^r$.

(b) The load that can be carried by a beam of breadth b, depth d, and length L is given by Cbd^3/L where C is a constant depending on the material of which the beam is made. If in a particular beam the breadth is increased by 1%, the depth is decreased by 2% and the length is decreased by 3%, use the method of binomial expansion to find by what per cent the load is affected.

D.T.C.

16 (a) Write down the expansion of $(1+x)^n$ as far as x^4. If $b = 1+c$ find an approximate value in terms of powers of c for the expression

$$2b^4 - 3b^3 - 5b$$

where powers of c above the second may be neglected.

(b) Use this result to find an approximation to the value of b near to unity given by the equation:

$$2b^4 - 3b^3 - 5b + 6.06 = 0$$

(b should be quoted to three places of decimals.)

U.E.I.

71   Miscellaneous exercises 4

17   The current i A, flowing at a time t s is given by the formula

$$i = \frac{V}{R}(1 - e^{-Rt/L})$$

If $V = 240$ V, $R = 8$ and $L = 0.32$, plot the current–time graph taking values of t at intervals of 0.02 V from $t = 0$ to $t = 0.2$ s.

18   State the series for $\log_e (1+x)$ and say for what values of x it is true.
In a certain chemical reaction

$$e^{0.05t} = \frac{1-x}{1-2x} \text{ when } 0 < x < \tfrac{1}{2}$$

Express t as a series in ascending powers of x up to the term in x^3.

19   (a) Solve $\sqrt{(x+6)} + \sqrt{(3x-5)} = \sqrt{(8x+1)}$.
(b) Evaluate d if f the stress in a plate is given by

$$f = \frac{P}{\pi t^2}\left\{\frac{4}{3}\log_e\left(\frac{D}{d}\right) + 1\right\}$$

when $f = 5180$, $P = 1200$, $t = 0.5$ and $D = 6$.

U.E.I.

20   Write down the series for $\log_e (1+x)$ and $\log_e (1-x)$ provided $|x| < 1$.
Prove that $\log_e \dfrac{1+x}{1-x} = 2\left(x + \dfrac{x^3}{3} + \dfrac{x^5}{5} + \ldots\right)$.

Show also that if $\dfrac{1+x}{1-x} = n$ then $x = \dfrac{n-1}{n+1}$, and hence find the value of $\log_e 1.5$ to four decimal places.
Why is it clear that if $x > 1$ the series can have no meaning?

D.T.C.

21   (a) Write down the series for e^x and hence find, to four decimal places, an approximation for $e^{0.02}(1-0.006)^{\frac{1}{4}}$.
(b) Using logarithms find S if $S = e^{kT}(1-T^{-k})$ given that $k = 0.273$ and $T = 0.106$.

U.L.C.I.

22   (a) On the same axes and to the same scales, plot the graphs of e^x and e^{-x} from $x = -2.5$ to $x = 2.5$.

(b) On the same axes construct the graph of $\dfrac{(e^x + e^{-x})}{2}$.

Read off this graph the values of x when the function $\dfrac{(e^x + e^{-x})}{2}$ has the value 1·25.

(c) Check the previous result by solving the equation $\dfrac{(e^x + e^{-x})}{2} = 1\cdot25$ as a quadratic in e^x.

U.E.I.

5 Further differentiation

5a Functions and limiting values

When two variable quantities x and y are so related that the value of one quantity y depends on the value of the other quantity x we say y is a *function* of x. We express this by writing $y = f(x)$.

For example if $y = 2x^2 - 3x + 1$ we may write

$$f(x) = 2x^2 - 3x + 1$$
$$\text{and} \quad f(2) = 2(2)^2 - 3(2) + 1 = 3$$
$$f(-1) = 2(-1)^2 - 3(-1) + 1 = 6$$
$$f(1) = 2(1)^2 - 3(1) + 1 = 0$$

Sometimes the value of a function has no meaning or is indeterminate for some value of x.

If $f(x) = \dfrac{x^2 - 4}{x - 2}$ then $f(2) = \dfrac{0}{0}$ which we cannot evaluate. In cases like this we often introduce the idea of a limit or limiting value.

Suppose we put $x = 2 + h$ in the expression.

$$\frac{x^2 - 4}{x - 2} = \frac{(2 + h)^2 - 4}{2 + h - 2} = \frac{4 + 4h + h^2 - 4}{h} = 4 + h$$

As x gets closer and closer to 2, then h approaches the value 0 and $\dfrac{x^2 - 4}{x - 2}$ gets closer and closer to the value 4.

We say the limit of $\dfrac{x^2 - 4}{x - 2}$ as x approaches 2 equals 4 and we write this

$$\lim_{x \to 2} \frac{x^2 - 4}{x - 2} = 4$$

Notice that the limit is not the same as $\dfrac{x^2 - 4}{x - 2}$ when $x = 2$ as this latter is indeterminate.

We say that $\dfrac{x^2 - 4}{x - 2}$ is discontinuous at $x = 2$.

A function of x is said to be continuous at the value $x = a$ if the value of the limit of $f(x)$ as x approaches a is equal to $f(a)$.

Example   Find the limit of $\dfrac{2x^2 - 5x - 3}{x - 3}$ when x approaches 3.

Let $x = 3 + h$ so that h approaches 0 when x approaches 3.

$$\text{Then } \lim_{x \to 3} \frac{2x^2 - 5x - 3}{x - 3} = \lim_{h \to 0} \frac{2(3 + h)^2 - 5(3 + h) - 3}{(3 + h) - 3}$$

$$= \lim_{h \to 0} \frac{7h + 2h^2}{h}$$

$$= \lim_{h \to 0} 7 + 2h$$

Putting $h = 0$ we see the required limit is 7.

Exercise 5a

1   If $f(x) = x^2 - 3x + 1$ find $f(1), f(2), f(0), f(x+1), f(x+h)$.

2   If $f(x) = \dfrac{x^2 + x - 2}{x - 1}$ find $f(2), f(0)$ and $f(1)$.

Find the following limits:

3   $\displaystyle\lim_{x \to 1} \frac{x^2 + x - 2}{x - 1}$
    4   $\displaystyle\lim_{x \to 2} \frac{2x^2 - x - 6}{x - 2}$

5   $\displaystyle\lim_{x \to 3} \frac{x^2 - 9}{x - 3}$
    6   $\displaystyle\lim_{x \to 0} \frac{2x^2 + 3x}{x}$

7   $\displaystyle\lim_{x \to 2} \frac{x^2 - x - 2}{2x^2 - 3x - 2}$
    8   $\displaystyle\lim_{x \to \infty} \frac{x}{2x + 1}$

Hint: write $\dfrac{x}{2x + 1} = \dfrac{1}{2 + \dfrac{1}{x}}$

9   $\displaystyle\lim_{h \to 0} \frac{(x + h)^2 - x^2}{h}$

10   If $f(x) = 3x^2$ evaluate $\displaystyle\lim_{h \to 0} \frac{f(x + h) - f(x)}{h}$.

11   If $f(x) = x^2 - 2x$ find $\displaystyle\lim_{h \to 0} \frac{f(x + h) - f(x)}{h}$.

12   If $f(x) = x^3 + 2x$ find $\lim\limits_{h \to 0} \dfrac{f(x+h) - f(x)}{h}$.

5b   The differential coefficient

In volume 1, chapter 7 the differential coefficient of $y = f(x)$ was found from a consideration of the fraction $\dfrac{\delta y}{\delta x}$ where δy represents the increase in y due to a small increase δx in x. The limiting value of $\dfrac{\delta y}{\delta x}$ as δx becomes smaller and smaller was called the *derivative* or *differential coefficient* of y and it may be said to be the instantaneous rate of change of y with respect to x. It is denoted by the symbol $\dfrac{dy}{dx}$ and is written

$$\frac{dy}{dx} = \lim\limits_{\delta x \to 0} \frac{\delta y}{\delta x}$$

Putting $y = f(x)$, then $y + \delta y = f(x + \delta x)$ and the fraction $\dfrac{\delta y}{\delta x}$ becomes $\dfrac{f(x + \delta x) - f(x)}{\delta x}$. Thus the derivative may be defined as

$$\frac{dy}{dx} = \lim\limits_{\delta x \to 0} \frac{f(x + \delta x) - f(x)}{\delta x}$$

and may be denoted by $\dfrac{d}{dx} f(x)$ or $f'(x)$.

This definition is useful in finding derivatives of functions of x from first principles, and it can save a few steps in the working.

We have been finding such limits in the previous section of this chapter, but using h instead of δx (indeed many people prefer h to δx) and defining the derivative of $f(x)$ as

$$\lim\limits_{h \to 0} \frac{f(x + h) - f(x)}{h}$$

In finding the differential coefficient of a function from first principles we may use any of the above definitions. Notice that unless $f(x)$ is continuous the limit could not exist.

Example 1   Find the differential coefficient of x^2.

Here $f(x) = x^2$.

In this example we will take δx as the small increment in x.

$$f(x+\delta x) = (x+\delta x)^2$$

$$\frac{f(x+\delta x)-f(x)}{\delta x} = \frac{(x+\delta x)^2 - x^2}{\delta x}$$

$$= \frac{x^2 + 2x\delta x + \delta x^2 - x^2}{\delta x}$$

$$= 2x + \delta x$$

Thus $\displaystyle\lim_{\delta x \to 0} \frac{f(x+\delta x)-f(x)}{\delta x} = 2x$ which is the required differential coefficient.

We may write $\dfrac{d}{dx} f(x) = 2x$ or $f'(x) = 2x$ or even $\dfrac{d}{dx}(x^2) = 2x$.

Example 2   Find the differential coefficient of x^n.

In this example we will take h and not δx as the small increment in x.

$$f(x) = x^n$$
$$f(x+h) = (x+h)^n$$
$$\frac{f(x+h)-f(\check{x})}{h} = \frac{(x+h)^n - x^n}{h}$$

$$= \frac{x^n\left[\left(1+\dfrac{h}{x}\right)^n - 1\right]}{h}$$

$$= \frac{x^n}{h}\left[1 + n\times\frac{h}{x} + \frac{n(n-1)}{2!}\frac{h^2}{x^2} + \ldots - 1\right]$$

by the binomial theorem.

$\left(\dfrac{h}{x} \text{ may be assumed } < 1 \text{ numerically as } h \to 0\right)$

$$= x^n\left[\frac{n}{x} + \text{terms containing } h \text{ and higher powers of } h\right]$$

Thus $\displaystyle\lim_{h \to 0} \frac{f(x+h)-f(x)}{h} = nx^{n-1}$

Hence if $f(x) = x^n$, $f'(x) = nx^{n-1}$

This result holds for all values of n, positive or negative, integral or fractional.

Thus if i $f(x) = x^3$ $f'(x) = 3x^2$

     ii $f(x) = \dfrac{1}{x^2} = x^{-2}$ $f'(x) = -2x^{-3} = -\dfrac{2}{x^3}$

iii $f(x) = \dfrac{1}{\sqrt{x}} = x^{-\frac{1}{2}}$ $f'(x) = -\dfrac{1}{2}x^{-\frac{3}{2}} = -\dfrac{1}{2\sqrt{x^3}}$

Example 3 Find the differential coefficients of sin x and cos x, where x is in radians.

i Let $\qquad f(x) = \sin x$

$\qquad\qquad f(x+\delta x) = \sin(x+\delta x)$

$$\frac{f(x+\delta x)-f(x)}{\delta x} = \frac{\sin(x+\delta x)-\sin x}{\delta x}$$

Remembering that $\sin - \sin = 2\cos(\tfrac{1}{2}\text{ sum})\sin(\tfrac{1}{2}\text{ difference})$

then $\qquad \dfrac{f(x+\delta x)-f(x)}{\delta x} = \dfrac{2\cos\left(x+\dfrac{\delta x}{2}\right)\sin\left(\dfrac{\delta x}{2}\right)}{\delta x}$

$$= \cos\left(x+\frac{\delta x}{2}\right)\frac{\sin\left(\dfrac{\delta x}{2}\right)}{\left(\dfrac{\delta x}{2}\right)}$$

Previously in volume 1, section 4d it was shown that

$$\lim_{\theta \to 0}\frac{\sin\theta}{\theta} = 1 \quad \text{so that} \quad \lim_{\delta x \to 0}\frac{\sin\left(\dfrac{\delta x}{2}\right)}{\dfrac{\delta x}{2}} = 1$$

Thus $\quad \displaystyle\lim_{\delta x \to 0}\frac{f(x+\delta x)-f(x)}{\delta x} = \cos x$

i.e. $\qquad \dfrac{d}{dx}(\sin x) = \cos x$

ii If $\qquad\qquad f(x) = \cos x$

then $\quad \dfrac{f(x+\delta x)-f(x)}{\delta x} = \dfrac{\cos(x+\delta x)-\cos x}{\delta x}$

Using the identity $\cos - \cos = -2\sin(\tfrac{1}{2}\text{ sum})\sin(\tfrac{1}{2}\text{ difference})$

$$\frac{f(x+\delta x)-f(x)}{\delta x} = \frac{-2\sin\left(x+\dfrac{\delta x}{2}\right)\sin\left(\dfrac{\delta x}{2}\right)}{\delta x}$$

$$= -\sin\left(x+\frac{\delta x}{2}\right)\frac{\sin\left(\dfrac{\delta x}{2}\right)}{\left(\dfrac{\delta x}{2}\right)}$$

Thus
$$\lim_{\delta x \to 0} \frac{f(x+\delta x)-f(x)}{\delta x} = -\sin x$$

i.e.
$$\frac{d}{dx}(\cos x) = -\sin x$$

These results are important and must be memorized.

5c Some useful theorems

These theorems, assumed in volume 1, are restated here with proofs.

I The derivative of a constant is zero

Let $y = C$ where C is a constant.
$y + \delta y = C$ i.e. C is unchanged by an increase in x.

$$\frac{\delta y}{\delta x} = \frac{C-C}{\delta x} = 0$$

$$\frac{dy}{dx} = \lim_{\delta x \to 0}(0) = 0$$

II The derivative of a constant times a function equals the constant times the derivative of the function

Let $y = C \times f(x)$
$y + \delta y = C \times f(x+\delta x)$

$$\frac{\delta y}{\delta x} = \frac{Cf(x+\delta x)-Cf(x)}{\delta x}$$

$$= C\frac{f(x+\delta x)-f(x)}{\delta x}$$

In the limit when $\delta x \to 0$

$$\frac{dy}{dx} = Cf'(x)$$

III The derivative of a sum of a number of functions of x equals the sum of their derivatives

Let $y = f_1(x)+f_2(x)+f_3(x)+ \ldots$
$y + \delta y = f_1(x+\delta x)+f_2(x+\delta x)+f_3(x+\delta x)+ \ldots$

$$\frac{\delta y}{\delta x} = \frac{f_1(x+\delta x)-f_1(x)}{\delta x}+\frac{f_2(x+\delta x)-f_2(x)}{\delta x}+\frac{f_3(x+\delta x)-f_3(x)}{\delta x}+ \ldots$$

Let δx approach 0 and in the limit

$$\frac{dy}{dx} = f_1'(x) + f_2'(x) + f_3'(x) + \ldots$$

Example   Find the derivatives with respect to x of the following:

i $3x^3 - 2x^2 - 3x + 4$ ii $\dfrac{x^7 + x^3}{x^5}$ iii $x^3(x^2 - 2x + 3)$

i $\dfrac{d}{dx}(3x^3 - 2x^2 - 3x + 4) = \dfrac{d}{dx}(3x^3) - \dfrac{d}{dx}(2x^2) - \dfrac{d}{dx}(3x) + \dfrac{d}{dx}(4)$

$$= 9x^2 - 4x - 3$$

ii $\dfrac{d}{dx}\left(\dfrac{x^7 + x^3}{x^5}\right) = \dfrac{d}{dx}\left(x^2 + \dfrac{1}{x^2}\right) = 2x - \dfrac{2}{x^3}$

iii $\dfrac{d}{dx}[x^3(x^2 - 2x + 3)] = \dfrac{d}{dx}(x^5 - 2x^4 + 3x^3) = 5x^4 - 8x^3 + 9x^2$

Exercise 5c

Differentiate from first principles

1   $5x$ 2 $2x^2 - 3x$ 3 $x^2 + 3$

4   $\sin 2x$ 5 $\dfrac{1}{x^2}$ 6 $\dfrac{1}{2x+1}$

In examples 7 to 20 write down the differential coefficients with respect to the variable in each.

7   $2x^{3 \cdot 5}$ 8 $\dfrac{1}{S^{2 \cdot 6}} - \dfrac{1}{S^{1 \cdot 5}}$ 9 $\sqrt[3]{t^2}$

10   $\cos\theta + \sin\theta$ 11 $5v^{1 \cdot 4}$ 12 $\dfrac{1}{p^{1 \cdot 2}}$

13   $2\sin t - 3\cos t$ 14 $\dfrac{1}{\sqrt{y^5}}$ 15 $2x - \sin x$

16   $3z^3 - 2\cos z$ 17 $x^2 - 4\sin x - 2\cos x$

18   $2(\cos\theta - \sin\theta)$ 19 $\dfrac{10}{v} + 3\sin v$ 20 $4t^{1 \cdot 4} - \dfrac{\cos t}{3} + 15$

21   (a) If $f(x) = x^3 - 2x^2 + x$, write down $f'(x)$.
     (b) If $f(\theta) = 2\sin\theta - \cos\theta$, write down $f'(\theta)$.

80   Further differentiation

At the point where $x = \frac{1}{2}$ $\quad \frac{dy}{dx} = 3 \times (\frac{1}{2})^2 = \frac{3}{4}$

Required gradient $= \frac{3}{4}$

Let the equation of the tangent be

$$y = mx + c$$

m stands for the gradient and has been found to be $\frac{3}{4}$. The tangent passes through the point on the curve whose abscissa is $\frac{1}{2}$.

At $x = \frac{1}{2}$, $y = x^3 = (\frac{1}{2})^3 = \frac{1}{8}$

Thus the point $(\frac{1}{2}, \frac{1}{8})$ must satisfy the equation $y = \frac{3}{4}x + c$
i.e. $\qquad\qquad\qquad\qquad\qquad\qquad \frac{1}{8} = \frac{3}{4} \times \frac{1}{2} + c$
Hence $\qquad\qquad\qquad\qquad c = \frac{1}{8} - \frac{3}{8} = -\frac{1}{4}$
Required equation is thus $\quad y = \frac{3}{4}x - \frac{1}{4}$
or $\qquad\qquad\qquad\qquad\qquad 4y = 3x - 1$

5e Successive differentiation

If $y = x^4$

$$\frac{dy}{dx} = 4x^3$$

$\frac{dy}{dx}$ is itself a function of x and may be differentiated.

Thus $\quad \dfrac{d}{dx}\left(\dfrac{dy}{dx}\right) = \dfrac{d}{dx}(4x^3) = 12x^2$

The expression $\dfrac{d}{dx}\left(\dfrac{dy}{dx}\right)$ is written $\dfrac{d^2y}{dx^2}$ and is called the *second differential coefficient* or *second derivative* of y with respect to x.

Similarly $\dfrac{d}{dx}\left(\dfrac{d^2y}{dx^2}\right)$ is written $\dfrac{d^3y}{dx^3}$ and is called the third differential co-efficient or third derivative of y with respect to x.

Example Find $\dfrac{d^2y}{dx^2}$ and $\dfrac{d^3y}{dx^3}$ if $y = 2x^5$

Answer $\quad \dfrac{dy}{dx} = 10x^4 \quad \dfrac{d^2y}{dx^2} = 40x^3 \quad \dfrac{d^3y}{dx^3} = 120x^2$

Exercise 5e

Write down the differential coefficients with respect to the variable in examples 1 to 10.

82 Further differentiation

22 If $f(x) = x^2 - 2x$ find $f'(x)$ from first principles.

5d Geometric interpretation of $\dfrac{dy}{dx}$

In volume 1 the value of $\dfrac{dy}{dx}$ at a point on the curve $y = f(x)$ was shown to give the gradient of the tangent at that point. The method was as follows.

Let P be a point on the curve with coordinates (x, y) and let Q be a nearby point on the graph with coordinates $(x + \delta x, y + \delta y)$.

Figure 7

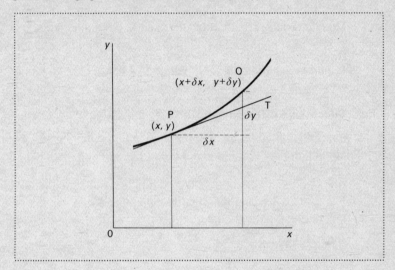

Then the gradient of the chord PQ $= \dfrac{\delta y}{\delta x} = \dfrac{f(x + \delta x) - f(x)}{\delta x}$

Let Q approach P so that δx approaches 0.

Then $\displaystyle\lim_{\delta x \to 0} \dfrac{f(x + \delta x) - f(x)}{\delta x}$ gives $f'(x)$ or $\dfrac{dy}{dx}$

and $\displaystyle\lim_{Q \to P}$ (gradient of chord PQ) gives the gradient of the tangent at P.

Example   Find the gradient of the tangent to the curve $y = x^3$ at the point where $x = \frac{1}{2}$, and obtain the equation of the tangent at this point.

$$y = x^3$$

$$\frac{dy}{dx} = 3x^2$$

1    $5 \sin x - 3 \cos x$

2    $2 + 3t + \dfrac{4}{t^2}$

3    $\dfrac{x^3}{3} - \dfrac{5x}{2} + 4 - \dfrac{3}{2x} + \dfrac{2}{x^2}$

4    $3t^2 - 4 - \dfrac{2}{t^2}$

5    $\sqrt{x} + \dfrac{1}{\sqrt{x}}$

6    $(\sqrt{x} + 2x)^2$

7    $\dfrac{x^3 - 1}{x^2}$

8    $\left(x + \dfrac{1}{x}\right)^3$

9    $\dfrac{1 - x^2 - x \cos x}{x}$

10    $(x^2 + 2)(x - 1)$

11   Find $\dfrac{d^2 y}{dx^2}$ if   (a) $y = x^3 + 2x - \dfrac{1}{x}$ (b) $y = (2x+1)(x^2 - 1)$

12   If $y = at^2 + bt$ where a and b are constants, prove that

$$t^2 \frac{d^2 y}{dt^2} - 2t \frac{dy}{dt} + 2y = 0$$

13   Verify that $\dfrac{d}{dx}\left(x \dfrac{dy}{dx}\right) = x \dfrac{d^2 y}{dx^2} + \dfrac{dy}{dx}$ when $y = x^3 + 2x$.

14   Find the equation of the tangent to $y = 2x - x^3$ at the point $(1, 1)$.

15   Find the equation of the tangent to the curve $y = (2+x)(3-2x)$ at the point on the curve where $x = 2$.

16   Find the points on the curve $y = x^3 + x^2 + x + 4$ where the gradient is 2.

5f   General methods of differentiation

Differential coefficients of more complicated functions than those dealt with above can be found from first principles, but the effort in some cases can be considerable. With the aid of the following theorems the differential coefficients of complicated functions can be found fairly quickly.

Differentiation of a product

Suppose $y = uv$ where u and v are both functions of x, for example
$$y = (x^3 - 2x^2) \sin x$$
Then $u = x^3 - 2x^2$ and $v = \sin x$

83   General methods of differentiation

Let x increase by δx and let the corresponding increases in u, v and y be δu, δv and δy, respectively.

Then $\quad y+\delta y = (u+\delta u)(v+\delta v)$
$$= uv+v\delta u+u\delta v+\delta u\delta v$$

and as $y = uv$ then $\delta y = v\delta u+u\delta v+\delta u\delta v$.

Dividing by $\delta x \quad \dfrac{\delta y}{\delta x} = v\dfrac{\delta u}{\delta x}+u\dfrac{\delta v}{\delta x}+\dfrac{\delta u}{\delta x}\times\dfrac{\delta v}{\delta x}\times\delta x$

As $\delta x \to 0$ $\dfrac{\delta y}{\delta x}, \dfrac{\delta u}{\delta x}$ and $\dfrac{\delta v}{\delta x}$ approach $\dfrac{dy}{dx}, \dfrac{du}{dx}$ and $\dfrac{dv}{dx}$.

So that $\quad \dfrac{dy}{dx} = v\dfrac{du}{dx}+u\dfrac{dv}{dx}$.

Example 1 If $y = (x^3-2x^2)\sin x$ find $\dfrac{dy}{dx}$.

$$\dfrac{dy}{dx} = (\sin x)\dfrac{d}{dx}(x^3-2x^2)+(x^3-2x^2)\dfrac{d}{dx}\sin x$$
$$= (3x^2-4x)\sin x+(x^3-2x^2)\cos x$$

Differentiation of a quotient

Suppose $y = \dfrac{u}{v}$ where u and v are both functions of x.

e.g. $\quad y = \dfrac{\cos x}{3x-x^3}$

Again let x increase by δx and let the corresponding increases in u, v and y be δu, δv and δy, respectively.

Then $\quad y+\delta y = \dfrac{u+\delta u}{v+\delta v}$

Hence $\quad \delta y = \dfrac{u+\delta u}{v+\delta v}-\dfrac{u}{v} = \dfrac{uv+v\delta u-uv-u\delta v}{(v+\delta v)\,v}$

$$\dfrac{\delta y}{\delta x} = \dfrac{v\dfrac{\delta u}{\delta x}-u\dfrac{\delta v}{\delta x}}{v^2+v\dfrac{\delta v}{\delta x}\times\delta x}$$

As $\delta x \to 0$, $\dfrac{\delta y}{\delta x}, \dfrac{\delta u}{\delta x}$ and $\dfrac{\delta v}{\delta x}$ approach $\dfrac{dy}{dx}, \dfrac{du}{dx}$ and $\dfrac{dv}{dx}$.

$$\frac{dy}{dx} = \frac{v\dfrac{du}{dx} - u\dfrac{dv}{dx}}{v^2}$$

Example 2 If $y = \dfrac{\cos x}{3x - x^3}$

$$\frac{dy}{dx} = \frac{(3x - x^3)\dfrac{d}{dx}\cos x - (\cos x)\dfrac{d}{dx}(3x - x^3)}{(3x - x^3)^2}$$

$$= \frac{(3x - x^3)(-\sin x) - (\cos x)(3 - 3x^2)}{(3x - x^3)^2}$$

$$= -\frac{(3x - x^3)\sin x + (3 - 3x^2)\cos x}{(3x - x^3)^2}$$

Note that the above rules can be extended to products and quotients containing more than two factors.

Example 3 If $y = \dfrac{x^2 \cos x}{\sin x}$

$$\frac{dy}{dx} = \frac{(\sin x)\dfrac{d}{dx}(x^2 \cos x) - x^2 \cos x\dfrac{d}{dx}\sin x}{\sin^2 x}$$

$$= \frac{(\cos x\, 2x - x^2 \sin x)\sin x - x^2 \cos x \cos x}{\sin^2 x}$$

$$= \frac{x \sin 2x - x^2}{\sin^2 x}$$

5g The differential coefficients of tan x, cot x, sec x, cosec x

These are all easily obtained from the differential coefficients of sin x and cos x using the formula for differentiating a quotient.

i
$$y = \tan x = \frac{\sin x}{\cos x} = \frac{u}{v}$$

$$\frac{dy}{dx} = \frac{(\cos x)\dfrac{d}{dx}\sin x - (\sin x)\dfrac{d}{dx}\cos x}{\cos^2 x}$$

$$= \frac{\cos x \cos x - \sin x\,(-\sin x)}{\cos^2 x}$$

$$\frac{dy}{dx} = \frac{\cos^2 x + \sin^2 x}{\cos^2 x}$$

$$= \frac{1}{\cos^2 x} = \sec^2 x$$

i.e. $\dfrac{d}{dx} \tan x = \sec^2 x$

ii $\qquad y = \sec x = \dfrac{1}{\cos x} = \dfrac{u}{v}$

$$\frac{dy}{dx} = \frac{(\cos x) \dfrac{d}{dx}(1) - 1 \dfrac{d}{dx} \cos x}{\cos^2 x}$$

$$= \frac{0 + \sin x}{\cos^2 x}$$

$$= \sec x \tan x$$

i.e. $\dfrac{d}{dx} \sec x = \sec x \tan x$

It is left as an exercise for the student to show

iii $\dfrac{d}{dx}(\cot x) = -\operatorname{cosec}^2 x$

iv $\dfrac{d}{dx}(\operatorname{cosec} x) = -\operatorname{cosec} x \cot x$

Exercise 5g

Differentiate the following products:

1 $x \cos x$

2 $(2x - 1) \sin x$

3 $(2x^4 - 3x)(x^3 + 2x - 1)$

4 $(\sqrt{x} + 1)(x^2 - x - 1)$

5 $x^3 \sin x$

6 $(x + 2) \tan x$

Differentiate the following quotients:

7 $\dfrac{\sin x}{x}$

8 $\dfrac{3x - 1}{x^2 + 1}$

9 $\dfrac{3x}{x - 1}$

10 $\dfrac{x^2 - 1}{\cos x}$

11 $\dfrac{\tan x}{x + 1}$

12 $\dfrac{3x^2 - x}{2x^2 - 3}$

Differentiate with respect to the variable:

13   $t \cos t$

14   $(t^2 + 2t - 1) \sin t$

15   $\dfrac{3 \sin x}{x^2 - 1}$

16   $\sin \theta \cos \theta$

17   $x^2 \tan x$

18   $\dfrac{x^3 - x}{\sin x}$

19   $\dfrac{x^3 + 6x^2 + 9x}{x + 1}$

20   $\dfrac{\tan \theta}{\theta - 1}$

21   $\dfrac{\sqrt{y+1}}{\sqrt{y-1}}$

22   $\dfrac{1}{z^2 + z}$

23   $(x^4 + 1) \tan x$

24   $x^3 (1 - \cos x)$

25   $(x^2 - 1) \sec x$

26   $x^2 \cot x$

27   $\dfrac{x \sin x}{(x - 1)}$

28   $\dfrac{(x + 1) \tan x}{x - x^2}$

Find $\dfrac{d^2 y}{dx^2}$ from the following:

29   $y = x \cos x$

30   $y = x^2 \sin x$

31   $y = \dfrac{\sin x}{x^2}$

32   $y = \dfrac{\cos x}{x}$

5h   Function of a function

We cannot as yet differentiate an expression such as $\sqrt{(x^2 + x + 3)}$.

If we put $u = x^2 + x + 3$ then $y = \sqrt{u}$ and y is a function of u, and u is a function of x, so that y is a function of a function of x.

In this case if x increases by δx, let δu be the change in u and δy the corresponding change in y.

By ordinary algebra $\dfrac{\delta y}{\delta x} = \dfrac{\delta y}{\delta u} \times \dfrac{\delta u}{\delta x}$

If δx now approaches 0 then $\delta u \to 0$ and in the limit

$$\dfrac{\delta y}{\delta x} \to \dfrac{dy}{dx}, \dfrac{\delta y}{\delta u} \to \dfrac{dy}{du} \quad \text{and} \quad \dfrac{\delta u}{\delta x} \to \dfrac{du}{dx}$$

Thus $\dfrac{dy}{dx} = \dfrac{dy}{du} \times \dfrac{du}{dx}$

In the example above $y = u^{\frac{1}{2}}$ and $\dfrac{dy}{du} = \frac{1}{2}u^{-\frac{1}{2}}$

$$u = x^2 + x + 3 \quad \text{and} \quad \dfrac{du}{dx} = 2x + 1$$

Thus $\qquad \dfrac{dy}{dx} = \frac{1}{2}u^{-\frac{1}{2}}(2x+1) = \dfrac{2x+1}{2\sqrt{(x^2+x+3)}}$

Example 1 Differentiate $(x^2 + 2x)^7$.

Let $u = x^2 + 2x$ then $y = u^7$

$$\dfrac{dy}{dx} = \dfrac{dy}{du} \times \dfrac{du}{dx} = 7u^6(2x+2)$$

and rewriting $x^2 + 2x$ for u, this gives $7(x^2 + 2x)^6(2x+2)$

Example 2 Differentiate $\sin(3x+1)$.

Let $(3x+1) = u$ and $y = \sin u$

$$\dfrac{du}{dx} = 3 \quad \text{and} \quad \dfrac{dy}{du} = \cos u$$

Thus $\qquad \dfrac{dy}{dx} = \dfrac{dy}{du} \times \dfrac{du}{dx} = (\cos u)\,3$
$$= 3\cos(3x+1)$$

The above examples are worked in full but with practice such differential coefficients can be written down at sight.

Example 3 Differentiate $\cos(2x^2 - 1)$.

$$\dfrac{d}{dx}\cos(2x^2 - 1) = \dfrac{d[\cos(2x^2-1)]}{d(2x^2-1)} \times \dfrac{d(2x^2-1)}{dx}$$
$$= [-\sin(2x^2-1)] \times 4x$$
$$= -4x\sin(2x^2-1)$$

Exercise 5h

In questions 1 to 5 find $\dfrac{dy}{du}$ and $\dfrac{du}{dx}$ and then multiply together to get $\dfrac{dy}{dx}$.

1 $y = u^4$ where $u = x^2 + x$ $\qquad$ i.e. $y = (x^2 + x)^4$

2 $y = 2u^6$ where $u = 3 - x^2$ $\qquad$ i.e. $y = 2(3 - x^2)^6$

3 $y = \sin u$ where $u = 3x^2 + 1$ $\quad$ i.e. $y = \sin(3x^2 + 1)$

4 $y = \tan u$ where $u = 4x$ $\qquad$ i.e. $y = \tan 4x$

5   $y = \cos u$ where $u = 2x^2 - 1$ i.e. $y = \cos(2x^2 - 1)$

In the following find $\dfrac{dy}{dx}$.

6   $y = (x^2 + x)^{11}$

7   $y = (1 - x^2)^8$

8   $y = (3x^2 - 4x + 5)^3$

9   $y = \sqrt{(3x^2 - 4x + 5)}$

10   $y = \dfrac{1}{x^2 - x - 1}$

11   $y = \sin(3x + 2)$

12   $y = \tan 2x$

13   $y = \sec 3x$

14   $y = (x^2 + x)^6$

15   $y = (x^2 + x)^{12}$

16   $y = (ax + b)^7$

17   $y = \tan(4x + 3)$

18   $y = (3x + 1)^7$

19   $y = \dfrac{1}{(3x + 1)^7}$

20   $y = \sqrt{(3x + 1)}$

21   $y = \sin^{10} x$

22   $y = \cos^{11} x$

23   $y = (x^2 + 1)^5$

24   $y = (\sin x + 1)^5$

25   $y = \dfrac{1}{(\sin x + 1)^2}$

26   $y = \sin^3 x$

27   $y = \cos^3 x$

28   $y = \cos 3x$

29   $y = \operatorname{cosec} 4x$

30   $y = \cot^4 x$

31   $y = \sqrt[3]{(x^3 + 8)}$

32   $y = \sin 4x$

33   $y = \cos(x^2 - 1)$

34   $y = \sqrt{(x + 1)}$

35   $y = \tan 3x$

36   $y = 4 \sec 2x$

37   $y = \dfrac{1}{\sqrt{(x + 2)}}$

38   $y = 2 \cot 3x$

39   $y = \tan(4x + 3)$

40   $y = (x - 1)^{10} \sin x$

41   $y = (x^2 - 1) \cos 3x$

89   Function of a function

42   $y = \dfrac{x}{\sin 4x}$ 43 $y = \dfrac{1}{\sin 3x}$

44   $y = \sqrt{(x^2+4)}\cos(2x+1)$ 45 $y = \dfrac{\sec 3x}{\sqrt{(2x-3)}}$

5i   To find $\dfrac{dy}{dx}$ from an implicit function

If x and y are given by a relationship such as $y^3+y^2+x^3-2x-7 = 0$ this is known as an 'implicit' function and y may be said to be an implicit function of x. In $y = x^3+2x-6$, y is an 'explicit' function of x.

We cannot easily rearrange an implicit function in the form $y = f(x)$ but we can still find a value for $\dfrac{dy}{dx}$ by the function of a function rule.

For example we differentiate y^3 with respect to x in the following way:
$\dfrac{d}{dx}(y^3) = \dfrac{d}{dy}(y^3)\times\dfrac{dy}{dx} = 3y^2\,\dfrac{dy}{dx}$. In this way we can differentiate an implicit function such as $y^3+y^2+x^3-2x-7 = 0$ with respect to x as it stands.

We get $\dfrac{d}{dx}(y^3)+\dfrac{d}{dx}(y^2)+\dfrac{d}{dx}(x^3)-\dfrac{d}{dx}(2x)-\dfrac{d}{dx}(7) = 0$

i.e. $3y^2\,\dfrac{dy}{dx}+2y\,\dfrac{dy}{dx}+3x^2-2 = 0$

Rearranging $\dfrac{dy}{dx}(3y^2+2y) = 2-3x^2$

From which we get $\dfrac{dy}{dx} = \dfrac{2-3x^2}{3y^2+2y}$

Notice that $\dfrac{dy}{dx}$ is given in terms of both x and y.

Example   If $x^2+y^2+5x-4y-3 = 0$ find $\dfrac{dy}{dx}$

Differentiating w.r.t. x we get

$2x+\dfrac{d}{dy}(y^2)\dfrac{dy}{dx}+5-4\dfrac{dy}{dx} = 0$

i.e. $2x+2y\,\dfrac{dy}{dx}+5-4\dfrac{dy}{dx} = 0$

i.e. $\dfrac{dy}{dx} = \dfrac{2x+5}{4-2y}$

5j To find $\dfrac{dy}{dx}$ when x and y are functions of a parameter

Sometimes x and y are given in terms of a third variable called a parameter. For example: $x = 2t + t^3$, $y = t^2 - 1$.

To find $\dfrac{dy}{dx}$ we proceed as follows:

Let t increase by δt, then x and y will increase by δx and δy respectively.

Algebraically $$\frac{\delta y}{\delta x} = \frac{\delta y}{\delta t} \div \frac{\delta x}{\delta t}$$

and proceeding to the limit $$\frac{dy}{dx} = \frac{dy}{dt} \div \frac{dx}{dt}$$

Thus in the above example $$\frac{dy}{dx} = 2t \div (2 + 3t^2)$$

$$= \frac{2t}{2 + 3t^2}$$

Note If in the rule $\dfrac{dy}{dx} = \dfrac{dy}{dt} \div \dfrac{dx}{dt}$ we put $t = y$ then $\dfrac{dy}{dt} = 1$ and we get the very useful result that

$$\frac{dy}{dx} = \frac{1}{\dfrac{dx}{dy}}$$

This last result is very useful in finding $\dfrac{dy}{dx}$ from certain functions which we call inverse functions.

Example 1 If $x = a \cos \phi$ and $y = a \sin \phi$ find $\dfrac{dy}{dx}$.

$$\frac{dy}{dx} = \frac{dy}{d\phi} \div \frac{dx}{d\phi} = \frac{a \cos \phi}{-a \sin \phi} = -\cot \phi$$

Example 2 Find the gradient of the curve $y^3 - 2y = 3x + 1$ at the point (1, 2).

Here we may use $\dfrac{dy}{dx} = \dfrac{1}{dx/dy}$ since $x = \dfrac{1}{3}(y^3 - 2y - 1)$.

$$\frac{dx}{dy} = \frac{1}{3}(3y^2 - 2) = \frac{1}{3}(3 \times 2^2 - 2) = \frac{10}{3} \text{ at the point (1, 2).}$$

Thus the gradient of the curve at (1, 2) $= \dfrac{dy}{dx}$ at the point (1, 2).

$$= \frac{3}{10}$$

Exercise 5j

In questions 1 to 9 find $\dfrac{dy}{dt}, \dfrac{dx}{dt}$ and $\dfrac{dy}{dx}$ in terms of t.

1   $x = t^2$
     $y = 2t$

2   $x = t^2 - 3$
     $y = 2 - t^2$

3   $x = 2t^3$
     $y = t^2$

4   $x = ct$
     $y = \dfrac{c}{t}$

5   $x = at^2$
     $y = 2at$

6   $x = t^2 + \dfrac{1}{t^2}$

     $y = t + \dfrac{1}{t}$

7   $x = a \cos t$
     $y = b \sin t$

8   $x = \sec t$
     $y = \tan t$

9   $x = a \cos^3 t$
     $y = a \sin^3 t$

In questions 10 to 13 find $\dfrac{dy}{dx}$ in terms of y.

10 $y^3 - y = x$

11 $(y+1)^3 = x$

12 $\sin y = x$

13 $y^2 + 2y - 4x + 2 = 0$

14 Find $\dfrac{dy}{dx}$ if (a) $x^2 + y^2 = 4$

        (b) $x^2 + y^2 + 2x - 6y - 4 = 0$ (c) $x^3 + 2xy + y^3 = 0$

5k Exponential and logarithmic functions

In Chapter 4 we met the number e as the sum

$$1 + 1 + \frac{1}{2!} + \frac{1}{3!} + \ldots \text{ to infinity} = 2{\cdot}718\,28\ldots$$

To differentiate the exponential function $y = e^x$

We may use the exponential theorem $e^x = 1 + x + \dfrac{x^2}{2!} + \dfrac{x^3}{3!} + \ldots$

Let $y = e^x$

$y + \delta y - y = e^{x + \delta x} - e^x = e^x (e^{\delta x} - 1)$

$$\frac{\delta y}{\delta x} = e^x \frac{\left[\left(1 + \delta x + \dfrac{(\delta x)^2}{2!} + \ldots\right) - 1\right]}{\delta x}$$

$$\frac{\delta y}{\delta x} = e^x \left(1 + \frac{\delta x}{2!} + \text{terms containing } (\delta x)^2 \text{ and higher powers}\right)$$

$$\frac{dy}{dx} = \lim_{\delta x \to 0} \frac{\delta y}{\delta x} = e^x$$

Thus e^x is a function whose gradient at any point is equal to the value of the function at that point.

Certain students may have omitted studying chapter 4 sections e and f dealing with the exponential and logarithmic functions, if their algebra syllabus does not include these functions. For them we give an independent treatment of the exponential function defining it in terms of the property that its gradient at any point on the graph of the function is equal to the value of the function itself at that point.

To differentiate a^x

Let $\quad y = a^x$

$$y + \delta y = a^{x + \delta x}$$

$$\delta y = a^{x + \delta x} - a^x = a^x(a^{\delta x} - 1)$$

and $\quad \dfrac{\delta y}{\delta x} = a^x \dfrac{a^{\delta x} - 1}{\delta x}$

Thus $\quad \dfrac{dy}{dx} = a^x \lim_{\delta x \to 0} \dfrac{a^{\delta x} - 1}{\delta x}$

Notice that this limit if it exists does not depend on x. This may be seen better if we write it $\lim\limits_{h \to 0} \dfrac{a^h - 1}{h}$. This limit is a function of a and we will denote it by the letter M.

Thus $\quad \dfrac{dy}{dx} = Ma^x$

When $x = 0$, $\dfrac{dy}{dx} = Ma^0 = M$ which makes M equal to the slope of the tangent to $y = a^x$ at the point where $x = 0$, i.e. at the point where the graph crosses the y-axis.

Thus if α is the angle to the horizontal that the tangent to the curve makes at the point where it crosses the y-axis then $\dfrac{d}{dx}(a^x) = a^x \tan \alpha$.

By plotting graphs of $y = a^x$ for different values of a we can find the particular value of a which makes $\tan \alpha = 1$. Denoting this value by the letter e we have found the function e^x such that $\dfrac{d}{dx}(e^x) = e^x$.

From the example below e will be seen to be between 2 and 3. By plotting $y = a^x$ for values of a between 2 and 3 e can be found to be 2·7. A more accurate value is $e = 2·718\ldots$.

Example Draw the graphs of the functions 1^x, 2^x, 3^x, $(\frac{1}{2})^x$ and $(\frac{1}{3})^x$ for values of x: $-3, -2, -1, 0, 1, 2, 3$. Measure the gradient of each at the point $x = 0$, $y = 1$.

x	-3	-2	-1	0	1	2	3
1^x	1	1	1	1	1	1	1
2^x	0·13	0·25	0·5	1	2	4	8
3^x	0·04	0·11	0·33	1	3	9	27
$(\frac{1}{2})^x$	8	4	2	1	0·5	0·25	0·13
$(\frac{1}{3})^x$	27	9	3	1	0·33	0·11	0·04

Figure 8

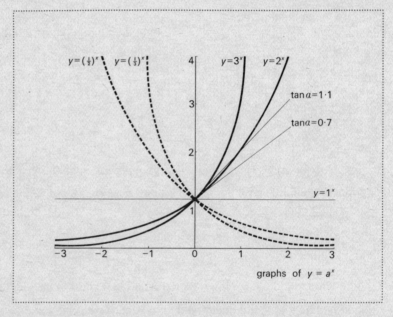

graphs of $y = a^x$

For the graph $y = a^x$ when $a = 2$, $\tan \alpha = 0·7$

when $a = 3$, $\tan \alpha = 1·1$

If we next plot $y = 2·5^x$ we will find $\tan \alpha = 0·9$ for this value of a, while $y = 2·7^x$ gives the slope $\tan \alpha = 0·99$.

Thus we find a function e^x such that

$$\frac{d}{dx}(e^x) = e^x \quad \text{where } e = 2·7 \text{ to two significant figures.}$$

The series for e^x

If we assume that e^x can be expanded as an infinite series in ascending powers of x we can obtain the series from the definition $\dfrac{d}{dx}(e^x) = e^x$ as follows:

Let
$$e^x \equiv a_0 + a_1 x + a_2 \frac{x^2}{2!} + a_3 \frac{x^3}{3!} + \ldots \tag{5.1}$$

Differentiating both sides

$$\frac{d}{dx}(e^x) = e^x = a_1 + a_2 x + a_3 \frac{x^2}{2!} + a_4 \frac{x^3}{3!} + \ldots \tag{5.2}$$

Equating coefficients in equation (5.1) and (5.2)

$a_0 = a_1, a_1 = a_2, a_2 = a_3, a_3 = a_4$, etc.

Putting $x = 0$ in (5.1) gives

$$e^0 = 1 = a_0 + a_1 \, 0 + a_2 \frac{0^2}{2!} + \ldots = a_0$$

Hence $\quad 1 = a_0 = a_1 = a_2 = a_3 = a_4 = \ldots$

Thus $\quad e^x = 1 + x + \dfrac{x^2}{2!} + \dfrac{x^3}{3!} + \dfrac{x^4}{4!} + \ldots$

To find the differential coefficient of $\log_e x$

If a number N can be written as $N = e^x$ then by our definition of logarithms x is the logarithm of N to base e. Logarithms to base e are known as hyperbolic or natural logarithms.

To find the derivative of $\log_e x$ we write

$$\log_e x = y$$

Then $\quad x = e^y \quad$ and differentiating with respect to y

$$\frac{dx}{dy} = e^y = x$$

Hence $\quad \dfrac{dy}{dx} = \dfrac{1}{x}$

Thus $\quad \dfrac{d}{dx} \log_e x = \dfrac{1}{x}$

When we return to integration this result supplies the missing term in the integration of x^n.

To find the differential coefficient of a^x

We must write a^x in the form e^z.

If $\qquad e^z = a^x \qquad$ taking logarithms to base e of both sides

$$z = \log a^x = x \log a$$

and so $\quad a^x = e^{x \log a}$

Differentiating by the rule for a function of a function

$$\frac{d}{dx} a^x = \frac{d}{d(x \log a)} e^{x \log a} \times \frac{d(x \log a)}{dx}$$

$$= e^{x \log a} \log a$$

$$= a^x \log_e a$$

To find the differential coefficient of $\log_a x$

Since $\quad \log_a x = \dfrac{\log_e x}{\log_e a}$

$$\frac{d}{dx}(\log_a x) = \frac{1}{\log_e a} \times \frac{d}{dx}(\log_e x) = \frac{1}{x \log_e a}$$

Here are two examples of this type:

$$\frac{d}{dx}(10^x) = 10^x \log_e 10 = 2{\cdot}3026 \times 10^x$$

$$\frac{d}{dx}(\log_{10} x) = \frac{1}{x \log_e 10} = \frac{1}{2{\cdot}3026 x}$$

Example 1 Differentiate i e^{ax} ii e^{x^2-1} iii $\log(ax+b)$ iv $\log \sin x$ v 2^x

i $\dfrac{d}{dx} e^{ax} = \dfrac{d}{d(ax)} e^{ax} \times \dfrac{d(ax)}{dx}$

$\qquad = e^{ax} \times a = ae^{ax}$

ii $\dfrac{d}{dx} e^{x^2-1} = \dfrac{d}{d(x^2-1)} e^{x^2-1} \times \dfrac{d}{dx}(x^2-1)$

$\qquad = e^{x^2-1} \times 2x = 2xe^{x^2-1}$

iii $\dfrac{d}{dx} \log(ax+b) = \dfrac{d}{d(ax+b)} \log(ax+b) \times \dfrac{d}{dx}(ax+b)$

$\qquad = \dfrac{1}{ax+b} \times a = \dfrac{a}{ax+b}$

iv $\dfrac{d}{dx}\log\sin x = \dfrac{d}{d(\sin x)}\log\sin x \times \dfrac{d}{dx}\sin x$

$$= \dfrac{1}{\sin x} \times \cos x = \cot x$$

v $\dfrac{d}{dx}(2^x) = 2^x \log_e 2 = 0.6931 \times 2^x$

Exercise 5k

Differentiate with respect to x

1 e^{3x}	2 $e^x + e^{-x}$	3 e^{x^2}
4 $\log(2x+3)$	5 $\log\cos x$	6 $\log(2x^2-1)$
7 $\log(1-x)$	8 $e^x \sin x$	9 4^x
10 $\dfrac{1}{e^{2x-1}}$	11 xe^x	12 $e^{2x}\cos 3x$
13 $x^2 \log x$	14 $e^{2x}\log\tan x$	15 $\log\dfrac{2-x}{2+x}$
16 $\dfrac{e^{ax}}{\sin bx}$	17 $\dfrac{x^2}{e^{2x}}$	18 $\dfrac{\log x}{e^{2x}}$
19 $x^n e^{ax}$	20 $xe^{\tan x}$	21 $e^{\cos x + \sin x}$
22 $\log(\sin x - \cos x)$		

5l Logarithmic differentiation

In some cases differentiation is made much easier if logarithms of the function are taken before differentiating. The logarithms are taken to base e and unless anything is said to the contrary $\log x$ is taken to be $\log_e x$ in the calculus.

Example 1 Find $\dfrac{dy}{dx}$ if $y = \dfrac{(x+1)(x-2)}{(x-3)(x-4)^2}$.

Taking logarithms, $\log y = \log(x+1) + \log(x-2) - \log(x-3) - 2\log(x-4)$.

Differentiating with respect to x and treating $\log y$ as a function of y where y is a function of x we get

$$\frac{1}{y} \times \frac{dy}{dx} = \frac{1}{x+1} + \frac{1}{x-2} - \frac{1}{x-3} - \frac{2}{x-4}$$

$$\frac{dy}{dx} = \frac{(x+1)(x-2)}{(x-3)(x-4)^2} \left[\frac{1}{x+1} + \frac{1}{x-2} - \frac{1}{x-3} - \frac{2}{x-4} \right]$$

Example 2   Find $\dfrac{dy}{dx}$ if $y = x^x$.

Taking logarithms, $\log y = x \log x$.

Differentiating with respect to x

$$\frac{1}{y} \frac{dy}{dx} = \log x + x \times \frac{1}{x}$$

$$\frac{dy}{dx} = x^x (\log x + 1)$$

Exercise 5l

Differentiate the following with respect to x.

1 $\dfrac{(x+2)(x-4)}{(x-1)(x-2)}$

2 $\dfrac{(x-1)^3}{(x+1)^2(x-2)}$

3 $\dfrac{(x-1)^{\frac{1}{2}}}{(x+1)^{\frac{1}{3}}(x+2)^{\frac{1}{3}}}$

4 $(x+1)^x$

5 $\sqrt[x]{(1+x)}$

6 x^{2x-1}

Miscellaneous exercises 5

1 Find $\dfrac{dy}{dx}$ given that

(a) $y = \dfrac{x^2+1}{x-1}$

(b) $y = \log_e(1+2x^2)$

(c) $y = 2 \sin 2x \sin x$

(d) $y = 3e^{\frac{1}{2}x^2}$ *U.L.C.I.*

2 (a) Differentiate with respect to x

i $\dfrac{x}{x^2-1}$

ii $e^x \sin x$

iii $\log_e(x^2-1)$

(b) Find the gradient of the tangent and the gradient of the normal to the curve $y = x^3 - 4x + 3$ at the point $(2, 3)$.

98   Further differentiation

3   (a) Differentiate with respect to x

   i $(2x^2-3)^{\frac{1}{2}}$ ii $\dfrac{x-1}{2x-3}$ iii $\sin^3 5x$

   (b) If $s = e^{-2t} \sin 3t$ show that

   $$\dfrac{d^2s}{dt^2}+4\dfrac{ds}{dt}+13s = 0$$

   D.T.C.

4   (a) Working from first principles derive the differential coefficient of $\sin 2\theta$ with respect to θ.

   (b) Differentiate with respect to x each of the following and simplify the results where possible.

   i $\dfrac{1}{\sqrt{(4-2x)}}$ ii $e^{-3x}\sin x$ iii $\dfrac{\sin 3x}{\cos^3 x}$ iv $(x^2-1)\log_e (2x+2)$

   D.T.C.

5   (a) If $y = uv$ where both u and v are functions of x deduce from first principles an expression for $\dfrac{dy}{dx}$.

   (b) Differentiate with respect to x

   i $\dfrac{1+3x^2}{1-2x^2}$ ii $\sin 3x \cos 3x$

   iii $\log_e \sqrt{(2x-5)}$ iv $e^{-ax} \cos x$ where a is constant.

6   (a) Differentiate with respect to x

   i $(x^3+2)^4$ ii $\cos (1-2x)$ iii $\dfrac{e^{2x}}{6x+3}$

   (b) If $y = \log_e (1+\sin x)$ prove that

   $$e^y \dfrac{d^2y}{dx^2} = -1$$

   D.T.C.

7   Differentiate with respect to x

   i $(1-x)\sqrt{(1+x)}$ ii $\dfrac{1+x^2}{e^{2x}}$ iii $\log_e (2x-1)$

   Show that $y = xe^{-x}$ satisfies the equation

   $$\dfrac{d^2y}{dx^2}+\dfrac{dy}{dx}+\dfrac{y}{x} = 0$$

   D.T.C.

99   Miscellaneous exercises 5

8 (a) Find $\dfrac{d^2y}{dx^2}$ in terms of x if $y(1-x) = 1+x$.

(b) If $z = \sqrt{(x^2+2x)}$, $u = x+1+\sqrt{(x^2+2x)}$ and
$y = [x+1+\sqrt{(x^2+2x)}]^{\frac{1}{2}}$
find $\dfrac{dz}{dx}$, and show that

i $\dfrac{du}{dx} = \dfrac{u}{\sqrt{(x^2+2x)}}$ ii $\dfrac{dy}{dx} = \dfrac{y}{2\sqrt{(x^2+2x)}}$

E.M.E.U.

9 (a) Find $\dfrac{dy}{dx}$ and express the results in the simplest form when

i $y = x^2 e^{2x}$ ii $y = \dfrac{(x^2+a^2)^3}{\sqrt{(x^2+a^2)}}$ iii $y = \log_e \cos ax$

(b) The equation of a cycloid formed by the rolling of a circle of radius a is given by the equations $x = a(\theta+\sin\theta)$; $y = a(1-\cos\theta)$, where θ is the angle of rotation of the circle.

Given $\dfrac{dy}{dx} = \dfrac{\dfrac{dy}{d\theta}}{\dfrac{dx}{d\theta}}$, show that when $\theta = 120°$ the value of $\dfrac{dy}{dx}$ is $\sqrt{3}$.

D.T.C.

10 (a) If $y = \sin x$, prove that $\dfrac{d^2y}{dx^2} = \sin(\pi+x)$.

(b) If $y = \dfrac{(x+1)^3(x-2)^2}{(x+2)^2\sqrt{(x+4)}}$ find $\dfrac{dy}{dx}$. (Take logs.)

(c) Find the equation of the tangent to the curve $y^3+3y^2 = 2x$ at the point (2, 1).

11 Find $\dfrac{dy}{dx}$ and $\dfrac{d^2y}{dx^2}$ when (a) $y = \sin x \sin 2x$

(b) $y = x^3 \sin x$

If $y = x^2(2+3\log_e x)$ show that

$x^2 \dfrac{d^2y}{dx^2} - 3x \dfrac{dy}{dx} + 4y = 0$

12 (a) Find the limiting value of $\dfrac{x^2+4x-5}{x-1}$ when $x = 1$.

(b) If $y = \dfrac{(x+1)^3(x-2)}{(x-3)^2(x+4)}$ find $\dfrac{dy}{dx}$ by taking logarithms of both sides and differentiating.

(c) Find the equation of the tangent to the curve $y^3 + 3y = x^2$ at the point (2, 1).

13   (a) Differentiate with respect to x

    i $\dfrac{3}{x^2-4}$ ii $x^2 \log x$ iii $\tan^2(3x-5)$

   (b) If $x = at(t+2)$ and $y = 2a(t+1)$, show that $\dfrac{d^2x}{dy^2} = \dfrac{1}{y}\dfrac{dx}{dy}$.

D.T.C.

14   (a) Differentiate with respect to x

    i $\dfrac{1-\cos x}{\sin x}$ ii $e^{2x}\sin 3x$ iii $\log\sec\left(5x+\dfrac{\pi}{4}\right)$

   (b) When a gas expands adiabatically, the pressure and volume are connected by the law $pv^\gamma = $ constant.

      Prove that $\dfrac{dp}{dt} = -\gamma\dfrac{p}{v}\dfrac{dv}{dt}$.

D.T.C.

15   (a) Differentiate with respect to x

    i $\log_e 2x^3$ ii $\dfrac{1-\sin x}{1+\sin x}$ iii $e^{\frac{1}{2}x}\tan x$

   (b) Given that $\dfrac{d^2y}{dx^2} = \dfrac{-2}{x^2}$ find y in terms of x if

             $y = 0$ when $x = 1$

and $\dfrac{dy}{dx} = 3$ when $x = 1$

U.E.I.

16   (a) Differentiate with respect to x

    i $\sqrt{\dfrac{x}{1-x^2}}$ ii $e^{-x}\sin 2x$ iii $\log_e\tan 2x$

   (b) If $y = Ae^{3x} + Be^{-x}$ where A and B are constants, show that

$\dfrac{d^2y}{dx^2} - \dfrac{2dy}{dx} - 3y = 0$

N.C.T.E.C.

17 (a) Find $\dfrac{dy}{dx}$ when:

i $y = \sqrt{(1-x^2)}$ ii $y = \dfrac{2x-3}{1-x^2}$ iii $y = \log_e \tan x$

expressing each answer in its simplest form.

(b) If $x = A \cos nt + B \sin nt$, show that $\dfrac{d^2x}{dt^2} = -n^2x$.

(c) Find the gradient of the tangent to the curve $2x^3 - 9xy + 2y^3 = 0$ at the point (1, 2) on the curve.

D.T.C.

18 (a) Find $\dfrac{dy}{dx}$ when:

i $y = \dfrac{5-4x}{5x-4}$ ii $y = \log_e \sin \tfrac{1}{2}x$ iii $y = e^x \cos 2x$

(b) Show that $y = x + \sqrt{(x^2+1)}$ satisfies the equation $\dfrac{dy}{dx} = \dfrac{y}{\sqrt{(x^2+1)}}$.

D.T.C.

19 (a) Differentiate the following with respect to x:

i $\dfrac{1}{(3-2x)^2}$ ii $\dfrac{1-x^2}{1+x^2}$ iii $(x^2+2x+2)\,e^{-x}$

(b) Verify that at the point $\left(e, \dfrac{1}{e}\right)$ on the curve $y = \dfrac{\log x}{x}$

$\dfrac{dy}{dx} = 0$ and $\dfrac{d^2y}{dx^2}$ is negative.

(c) If $y = \log \tan 2x$, prove that $\dfrac{dy}{dx} = 4 \operatorname{cosec} 4x$.

E.M.E.U.

20 (a) Differentiate with respect to x

i $\log \dfrac{1}{x}$ ii $\sin\left(\dfrac{\pi}{6} - \dfrac{x}{2}\right)$ iii $\dfrac{x^2}{x^2+1}$

(b) If $e^{x+y} - 2x = 0$ show that $\dfrac{dy}{dx} = \dfrac{1}{x} - 1$.

(c) Show that $y = e^{3x} \cos 2x$ satisfies the equation

$$\dfrac{d^2y}{dx^2} - 3\dfrac{dy}{dx} + 4y + 6e^{3x} \sin 2x = 0$$

U.E.I.

6 Applications of differentiation

Chapter 8 of volume 1 dealt with applications of the differential calculus to problems involving rates of change, maxima and minima of functions, approximations, equations of tangents, etc. The more advanced rules for differentiation enable us to deal with problems previously too difficult to solve.

6a Rates of change

Example 1 A gas is compressed in a cylinder by a piston which moves so that the volume is decreasing by 2 m^3 s^{-1}. The pressure p kN m^{-2} and the volume v m^3 are connected by the formula $pv = $ a constant. If initially $p = 15$ kN m^{-2} and $v = 1$ m^3, find the rate of increase of the pressure when the volume is i 0·8 m^3 ii 0·4 m^3.

From the initial values of p and v we have

$$pv = 15 \times 1 = 15$$

$$p = \frac{15}{v}$$

$$\frac{dp}{dv} = -\frac{15}{v^2}$$

Also $\dfrac{dv}{dt} = -2$ m^3 s^{-1} (negative, since v is decreasing)

By the function of a function rule

$$\frac{dp}{dt} = \frac{dp}{dv} \times \frac{dv}{dt} = -\frac{15}{v^2} \times -2$$

$$= \frac{30}{v^2}$$

i When $v = 0·8$ $\quad \dfrac{dp}{dt} = \dfrac{30}{0·8^2} = \dfrac{30}{0·64} = 46·9$

The pressure is increasing at $46.9 \text{ kN m}^{-2}\text{s}^{-1}$.

ii When $v = 0.4$ $\quad \dfrac{dp}{dt} = \dfrac{30}{0.4^2} = 187.5 \text{ kN m}^{-2}\text{s}^{-1}$

Example 2   A connecting rod of length 1·5 m slides with its ends on two intersecting guides, one vertical and one horizontal. If one end of the rod is 90 cm from the point of intersection of the guides and is moving away horizontally at 1·2 m s^{-1}, find the velocity of the end on the other guide.

Figure 9

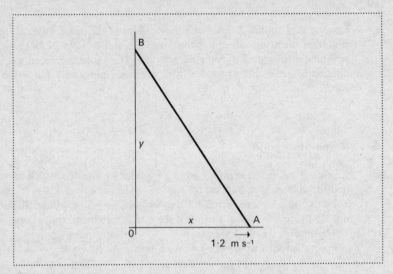

In fig. 9 AB represents the rod and O the intersection of the guides. Let $OA = x$ and $OB = y$.

Then $\quad x^2 + y^2 = 1.5^2$

The rate of change of x is given and we have to find the rate of change of y.
Differentiating $x^2 + y^2 = 1.5^2$ with respect to t

$$\frac{d}{dt}x^2 + \frac{d}{dt}y^2 = 0$$

i.e. $2x\dfrac{dx}{dt} + 2y\dfrac{dy}{dt} = 0$

hence $\dfrac{dy}{dt} = -\dfrac{x}{y}\dfrac{dx}{dt}$

$\dfrac{dx}{dt} = 1.2 \text{ m s}^{-1}$, and when $x = 90$ cm, $y = \sqrt{(150^2 - 90^2)} = 120$ cm i.e. $\dfrac{dy}{dt} = -\dfrac{90}{120} \times 1 \cdot 2 \text{ m s}^{-1} = -0 \cdot 9 \text{ m s}^{-1}$

The negative sign shows that y is decreasing.

6b Maximum and minimum values of a function

Figure 10

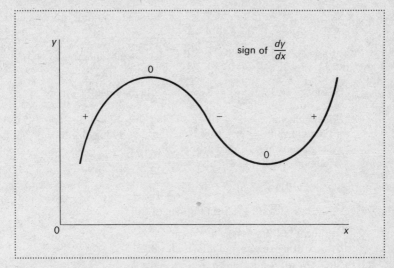

$\dfrac{dy}{dx} = 0$ at a maximum value and at a minimum value of y.

There are two methods of distinguishing between a maximum value and a minimum value.

The commoner test is to use the fact that $\dfrac{d^2y}{dx^2}$ is negative at a maximum and is positive at a minimum.

If the calculation of $\dfrac{d^2y}{dx^2}$ is difficult, a second test is available, using the fact that at a maximum $\dfrac{dy}{dx}$ changes sign from $+$ just before to $-$ just after, and at a minimum $\dfrac{dy}{dx}$ changes from $-$ just before to $+$ just after. See volume 1, section 8c.

Example 1 Find the maximum and minimum values of

i $y = x^2 - 10x + 12 \log_e x$
ii $y = \cos \theta \, (1 + \sin \theta)$ which occur within the range $\theta = 0$ to $\theta = \pi$.

i $y = x^2 - 10x + 12 \log_e x$

$$\frac{dy}{dx} = 2x - 10 + \frac{12}{x}$$

= 0 at maximum and minimum values.

$2x^2 - 10x + 12 = 0$

$2(x - 2)(x - 3) = 0$

i.e. $x = 2$ or 3

$$\frac{d^2y}{dx^2} = 2 - \frac{12}{x^2}$$

When $x = 2$ $\dfrac{d^2y}{dx^2} = 2 - \dfrac{12}{4} = -1$ (a maximum)

When $x = 3$ $\dfrac{d^2y}{dx^2} = 2 - \dfrac{12}{9} = \dfrac{2}{3}$ (a minimum)

Maximum value $= 2^2 - 10 \times 2 + 12 \log_e 2 = 12 \log_e 2 - 16 = -7{\cdot}683$
Minimum value $= 3^2 - 10 \times 3 + 12 \log_e 3 = 12 \log_e 3 - 21 = -7{\cdot}817$
Maximum value: $-7{\cdot}683$ at $x = 2$
Minimum value: $-7{\cdot}817$ at $x = 3$

ii $y = \cos \theta \, (1 + \sin \theta)$

$$\frac{dy}{d\theta} = \cos \theta \, (\cos \theta) + (1 + \sin \theta)(- \sin \theta)$$

$$= \cos^2 \theta - \sin \theta - \sin^2 \theta$$

= 0 at maximum and minimum values.

Thus $\cos^2 \theta - \sin \theta - \sin^2 \theta = 0$
Writing $\cos^2 \theta = 1 - \sin^2 \theta$ the equation becomes

$1 - \sin \theta - 2 \sin^2 \theta = 0$

$(1 - 2 \sin \theta)(1 + \sin \theta) = 0$

i.e. $\sin \theta = \frac{1}{2}$ or -1
i.e. $\theta = 30°, 150°$ or $270°$

Between 0 and π the required values are $\dfrac{\pi}{6}$ and $\dfrac{5\pi}{6}$.

$$\frac{dy}{d\theta} = \cos^2 \theta - \sin \theta - \sin^2 \theta$$

$$= \cos 2\theta - \sin \theta \quad (\text{since } \cos^2 \theta - \sin^2 \theta = \cos 2\theta)$$

$$\frac{d^2y}{d\theta^2} = -2 \sin 2\theta - \cos \theta$$

When $\theta = \dfrac{\pi}{6} = 30°; \dfrac{d^2y}{d\theta^2} = -2 \sin 60° - \cos 30°$

$$= -2 \times \frac{\sqrt{3}}{2} - \frac{\sqrt{3}}{2} = -\frac{3\sqrt{3}}{2}$$

This is negative which shows y is a maximum.

When $\theta = \dfrac{5\pi}{6} = 150°$; $\dfrac{d^2y}{d\theta^2} = -2\sin 300° - \cos 150°$

$$= -2\left(-\frac{\sqrt{3}}{2}\right) + \frac{\sqrt{3}}{2} = \frac{3\sqrt{3}}{2}$$

This is positive which shows y is a minimum.

When $\theta = \dfrac{\pi}{6}$, $y = \dfrac{\sqrt{3}}{2}(1 + \tfrac{1}{2}) = \dfrac{3\sqrt{3}}{4}$; a maximum value.

When $\theta = \dfrac{5\pi}{6}$, $y = -\dfrac{\sqrt{3}}{2}(1 + \tfrac{1}{2}) = -\dfrac{3\sqrt{3}}{4}$; a minimum value.

Example 2 Find the height and the radius of the cylinder of maximum volume which can be enclosed in a sphere of radius 6 cm.

Figure 11

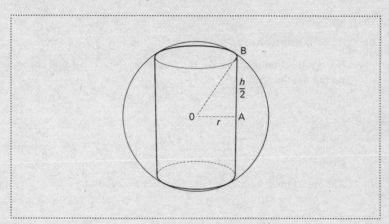

Let h be the height and r the radius of the cylinder. From fig. 11 we have
$OA^2 + AB^2 = OB^2$

i.e. $r^2 + \left(\dfrac{h}{2}\right)^2 = 6^2$

Volume of cylinder $V = \pi r^2 h = \pi h \left(36 - \dfrac{h^2}{4}\right)$.

We now have V as a function of h and for a maximum value of V, $\dfrac{dV}{dh} = 0$.

$$V = \pi\left(36h - \frac{h^3}{4}\right)$$

$$\frac{dV}{dh} = \pi\left(36 - \frac{3h^2}{4}\right) = 0$$

$$\frac{3h^2}{4} = 36$$

$$h^2 = 48, h = 4\sqrt{3}$$

$$\frac{d^2V}{dh^2} = \pi\left(-\frac{3h}{2}\right) \text{ which is negative}$$

V is a maximum when $h = 4\sqrt{3} = 6{\cdot}928$

$$r^2 = 6^2 - \left(\frac{h}{2}\right)^2 = 36 - 12 = 24$$

$$r = 2\sqrt{6} = 4{\cdot}899$$

Height of cylinder = 6·928 cm; radius = 4·899 cm.

6c Approximations

If y is a function of x then for the change in y due to a small change δx in x we can use the result

$$\delta y \simeq \frac{dy}{dx} \times \delta x$$

Example 1 $y = \sin x$. If x changes from 45° to 45° 3′ find the approximate change in y.

The angles must be changed from degrees to radians.

$$45° = \frac{\pi^c}{4}, 3' = \frac{3}{60} \times \frac{\pi}{180} \text{ rad}$$

$$\delta y \simeq \frac{d}{dx}(\sin x) \times \delta x$$

$$\simeq \cos x \times \delta x$$

$$\simeq \cos\frac{\pi}{4} \times \frac{3}{60} \times \frac{\pi}{180} = \frac{0{\cdot}7071 \times \pi}{3600} = 0{\cdot}000\,62$$

Example 2   Find the approximate error in calculating the area of a triangle in which two sides are 8 m and 10 m if the included angle is measured in error as 30° when it is actually 30° 12′.

As we have two sides and the included angle we use the formula

Area $A = \frac{1}{2}ab \sin C$

The angle C is the variable.

$$\delta A = \frac{dA}{dC} \times \delta C = \tfrac{1}{2}ab \cos C \times \delta C$$

$$\delta C = 12' = \frac{12}{60} \times \frac{\pi}{180} = \frac{\pi}{900} \text{ rad}$$

$$\delta A \simeq \tfrac{1}{2} \times 8 \times 10 \cos 30 \times \frac{\pi}{900} = 0\cdot121 \text{ m}^2$$

The calculated area will be approximately $0\cdot121$ m² too small.

Exercise 6c

1 The distance s m moved by a body after t s is given by $s = 20t - 2t^2 - \tfrac{1}{6}t^3$.

Find (a) its velocity and acceleration after 2 s,
　　(b) the time when the velocity is zero,
　　(c) the distance the body travels before first coming to rest.

2 The angle θ in radians through which a wheel has turned after t s is given by

$$\theta = 16t - \frac{t^2}{2}$$

　Find the angular velocity at the start and after 2 s. Show that the wheel has a constant retardation and find the time at which it first comes to rest.
　How many revolutions will it have made in this time?

3 A stone dropped into a still pond sends out circular ripples. If the ripple moves out from the centre at $0\cdot5$ m s^{-1} find the rate at which the disturbed area is increasing when the radius of the outermost ripple is (a) 4 m, (b) 6 m.

4 At a certain port, t hours after high water, the height h m of the tide above a fixed datum level is given by

$$h = 3 + 2 \cos \frac{\pi t}{6}$$

　At what rate in millimetres per second is the tide falling two hours after high water?

5 Water leaks from a pipe at $1\cdot2$ cm³ s^{-1} and forms a circular pool of $\tfrac{1}{4}$ cm constant depth. Find the rate at which the radius is increasing after 4 s.

6 A ladder of length 5 m rests against a vertical wall with the bottom end on horizontal ground. If the foot of the ladder is 3 m from the wall and moving away at 20 cm s^{-1}, find the rate at which the top end is sliding down the wall.

7   Water is running into an inverted cone of semi-vertical angle 45° at a steady rate of 10 dm³ s⁻¹. Find the rate at which the level is rising when the depth of water is (a) 60 cm, (b) 100 cm.

8   (a) Find the turning points of the curve $y = x^3 + x^2 - x + 1$, and distinguish between them.
     (b) Find the radius and height of a cylinder of maximum volume which can be inscribed in a sphere of radius 3 units.

9   (a) Find the maximum and minimum values of the function

$$y = x^3 - 6x^2 + 9x + 1$$

     (b) The velocity V of waves of length x in a canal is given by $V^2 = \dfrac{gx}{2\pi} + \dfrac{2\pi T}{\rho x}$ where T is the surface tension of the water and ρ is its density. Show that the velocity V is least when the wavelength $x = 2\sqrt{\dfrac{T}{\rho g}}$ and find the least value of V.

10   Find the maximum and the minimum values of:

     (a) $y = \dfrac{x^2 - 2x - 5}{2x + 3}$

     (b) $y = \cos 2x + \sin x$ from $x = 0$ to $x = \dfrac{\pi}{2}$ inclusive.

11   A cylinder is inscribed in a given cone. Prove that the maximum volume of the cylinder is $\frac{4}{9}$ that of the cone.

12   The torque T on a crankshaft is given by $T = 6 + 2\cdot4 \sin 2\theta - 3\cdot2 \cos 2\theta$.
     Find (a) the values of θ between 0° and 180° at which T is a maximum or minimum, (b) the corresponding values of T.

13   The stiffness of a rectangular beam whose breadth of section is x and depth is y is proportional to xy^3.
     Find the dimensions of the stiffest beam whose perimeter is 36 units.

14   Find the maximum and minimum values of:

     (a) $12 \log_e x - x^2 + 2x$
     (b) $\cos \theta (1 + \sin \theta)$ for values of θ between 0 and π.

15   A solid circular cylinder of radius r has a volume V. Show that its total surface area is $\dfrac{2V}{r} + 2\pi r^2$.

     Find the height and diameter of the cylinder of volume 200 cm³ which has the least surface area.

16   Find the approximate error in calculating the area of a triangle in which the two sides are 6 m and 10 m and the included angle is 0·87 rad, if there is an error of 1% in the value of the angle.

17   Find the approximate change in y if x changes from $40°$ to $40° 6'$ when
(a) $y = \cos x$
(b) $y = \tan 2x$

18   Find the approximate change in $\sqrt{x}$ when x increases by δx. Hence find the corresponding percentage change in $\sqrt{x}$.
Calculate the percentage error if we replace $\sqrt{898}$ by $\sqrt{900}$.

19   For a triangle, side c is calculated from the formula

$$c^2 = a^2 + b^2 - 2ab \cos C$$

where $a = 10$, $b = 8$, and angle $C = 60°$.
If there is an error of 2% in the value of a, the other terms being correct, find the approximate error in c.

20   For a converging lens of focal length 20 cm, the distance x cm of the object from the lens and the distance y cm of the image from the lens are connected by the relationship $(x-20)(y-20) = 400$.
If an object 50 cm from the lens is shifted 4 mm towards the lens, find the approximate shift in the position of the image.

6d   Tangents and normals to plane curves

The following results were obtained in volume 1, chapters 5 and 8.
The equation $y = mx + c$ represents a straight line whose gradient is m and which makes an intercept c on the y-axis. See fig. 12.

Figure 12

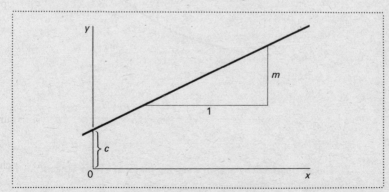

The equation $y - y_1 = m(x - x_1)$ represents a straight line passing through the point (x_1, y_1) with gradient m. See fig. 13.

Figure
13

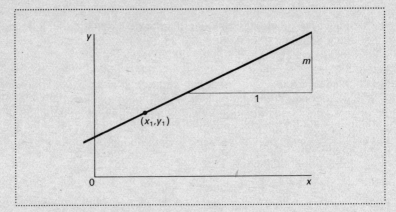

The equation $y - y_1 = \left[\dfrac{dy}{dx}\right]_1 (x - x_1)$, where $\left[\dfrac{dy}{dx}\right]_1$ is the value of $\dfrac{dy}{dx}$ at the point (x_1, y_1) is the equation of the tangent to the curve $y = f(x)$ at the point (x_1, y_1).

Figure
14

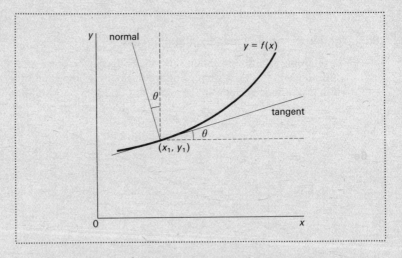

This is obtained by substituting the gradient of the tangent at the point (x_1, y_1) for m in the equation $y - y_1 = m(x - x_1)$.

112   Applications of differentiation

This gradient equals the value of $\dfrac{dy}{dx}$ for the curve at the point (x_1, y_1). See fig. 14.

Since the normal at the point (x_1, y_1) is perpendicular to the tangent to the curve at this point the gradient of the normal is given by

$$m = -\dfrac{1}{\left[\dfrac{dy}{dx}\right]_1}$$

When two lines are perpendicular the product of their gradients $m_1 m_2 = -1$.

Thus the equation of the normal at (x_1, y_1) is

$$y - y_1 = -\dfrac{1}{\left[\dfrac{dy}{dx}\right]_1}(x - x_1)$$

Example Find the equations of the tangent and normal at the point where $x = 2$ on the curve $y = x^2 + 3x$.

When $x = 2$, $y = 2^2 + 3 \times 2 = 10$

$$\dfrac{dy}{dx} = 2x + 3$$

At the point where $x = 2$ $\dfrac{dy}{dx} = 2 \times 2 + 3 = 7$

Thus equation of tangent is given by

$y - 10 = 7(x - 2)$
i.e. $y = 7x - 4$

Equation of normal is

$\qquad y - 10 = -\tfrac{1}{7}(x - 2)$
i.e. $7y - 70 = -x + 2$
i.e. $x + 7y = 72$

6e Newton's method of approximation to the root of an equation

This states that if $x = a$ is an approximate solution of $f(x) = 0$ then

$a - \dfrac{f(a)}{f'(a)}$ is a better approximation.

This result can be used repeatedly to give a step-by-step closer approximation to the root. It is known as the Newton (or Newton–Raphson) approximation formula, and can be obtained as follows.

113 Newton's method of approximation

Figure 15

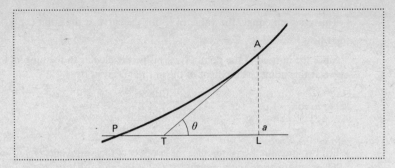

Let the graph of $y = f(x)$ cut the x-axis at P. See fig. 15.

Suppose A is the point on the curve whose abscissa is a and let the tangent at A cut the x-axis at T.

Then the value of x at T is a better approximation to the real root at P than is a, the value of x at point A.

$$AL = f(a) \qquad TL = \frac{AL}{\tan \theta}$$

But $\tan \theta$ = slope of the tangent at A

$$= \frac{dy}{dx} \text{ at A}$$

$$= f'(a)$$

Value of x at T is $a - TL = a - \dfrac{f(a)}{f'(a)}$.

Example 1 Given $x = 2$ as an approximate value of a root of $x^3 = 3x + 5$, find a closer approximation using Newton's method.

$$f(x) = x^3 - 3x - 5 \qquad f(2) = 8 - 6 - 5 = -3$$
$$f'(x) = 3x^2 - 3 \qquad f'(2) = 12 - 3 = 9$$

Taking $x = 2$ as a first approximation, a second and closer approximation

is $2 - \dfrac{f(2)}{f'(2)} = 2 - \dfrac{-3}{9} = 2\frac{1}{3}$

Using the method once more

$$f(2\tfrac{1}{3}) = (\tfrac{7}{3})^3 - 3(\tfrac{7}{3}) - 5 = 12 \cdot 70 - 7 - 5 = 0 \cdot 70$$
$$f'(2\tfrac{1}{3}) = 3(\tfrac{7}{3})^2 - 3 \qquad = 16\tfrac{1}{3} - 3 \qquad = 13 \cdot 3$$

Thus a third and even closer approximation is

$$2 \cdot 333 - \frac{0 \cdot 70}{13 \cdot 3} = 2 \cdot 333 - 0 \cdot 053$$

$$= 2 \cdot 28$$

Example 2 Make a table of values of $y = 2x^3 - 15x + 3$ between $x = 0$ and $x = 3$ and show that the equation $2x^3 - 15x + 3 = 0$ has two roots in this range. Find the value of the larger root correct to three significant figures.

$f(x) = 2x^3 - 15x + 3$
$f(0) = 3$
$f(1) = 2 \times 1^3 - 15 \times 1 + 3 = -10$
$f(2) = 2 \times 2^3 - 15 \times 2 + 3 = -11$
$f(3) = 2 \times 3^3 - 15 \times 3 + 3 = 12$

Setting these values out in a table we get

x	0	1	2	3
y	3	-10	-11	12

and the graph is shown in fig. 16.

Figure 16

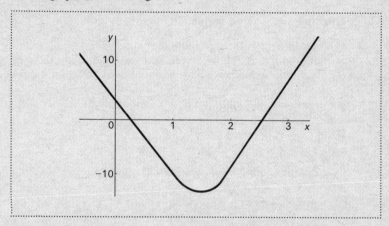

We see that the graph of $y = 2x^3 - 15x + 3$ crosses the x-axis between $x = 0$ and $x = 1$ and again between $x = 2$ and $x = 3$. Hence the equation has a root between 0 and 1 and one between 2 and 3. As a first approximation we take the smaller root as 0·2 and the larger root as 2·5.

We use Newton's formula for a closer approximation to the larger root.

$f(x) = 2x^3 - 15x + 3$
$f(2·5) = 2(2·5)^3 - 15(2·5) + 3$
$\quad\quad = -3·25$
$f'(x) = 6x^2 - 15$
$f'(2·5) = 6(2·5)^2 - 15 = 22·5$

The second approximation is $2·5 - \dfrac{-3·25}{22·5} = 2·5 + 0·14$

$$= 2·64$$

Continuing with the method

$$f(2{\cdot}64) = 2(2{\cdot}64)^3 - 15(2{\cdot}64) + 3 = 0{\cdot}2$$
$$f'(2{\cdot}64) = 6(2{\cdot}64)^2 - 15 \qquad\quad = 26{\cdot}82$$

The third approximation is $2{\cdot}64 - \dfrac{0{\cdot}2}{26{\cdot}82} = 2{\cdot}64 - 0{\cdot}007$

$$= 2{\cdot}63$$

Continuing the method once more gives the answer

$$2{\cdot}63 - \frac{f(2{\cdot}63)}{f'(2{\cdot}63)} = 2{\cdot}63 - \frac{-0{\cdot}07}{26{\cdot}5}$$
$$= 2{\cdot}63 + 0{\cdot}002$$
$$= 2{\cdot}63 \text{ to three significant figures.}$$

6f The sine and cosine series

In chapter 4 we gave the expansions of e^x and $\log_e (1+x)$ as infinite series in ascending powers of x. If we assume that $\sin x$ and $\cos x$ can be expressed as infinite series in ascending powers of x they may be obtained as follows:

$$\text{Let}\quad \sin x \equiv a_0 + a_1 x + a_2 \frac{x^2}{2!} + a_3 \frac{x^3}{3!} + a_4 \frac{x^4}{4!} + a_5 \frac{x^5}{5!} + \dots \tag{6.1}$$

When $x = 0$, $\sin x = 0$
$$a_0 = 0$$

Differentiating both sides of equation (6.1) with respect to x we get

$$\cos x \equiv a_1 + a_2 x + a_3 \frac{x^2}{2!} + a_4 \frac{x^3}{3!} + a_5 \frac{x^4}{4!} + \dots \tag{6.2}$$

When $x = 0$, $\cos x = 1$
$$a_1 = 1$$

Differentiating both sides of equation (6.2) with respect to x we get

$$-\sin x \equiv a_2 + a_3 x + a_4 \frac{x^2}{2!} + a_5 \frac{x^3}{3!} + \dots \tag{6.3}$$

Substitute in equation (6.3) the value of $\sin x$ in equation (6.1).

$$-a_0 - a_1 x - a_2 \frac{x^2}{2!} - a_3 \frac{x^3}{3!} - \dots \equiv a_2 + a_3 x + a_4 \frac{x^2}{2!} + a_5 \frac{x^3}{3!} + \dots$$

Equating coefficients gives

$$a_2 = -a_0 = 0 \qquad a_3 = -a_1 = -1$$
$$a_4 = -a_2 = 0 \qquad a_5 = -a_3 = 1 \quad \text{and so on.}$$

$$\sin x = x - \frac{x^3}{3!} + \frac{x^5}{5!} - \frac{x^7}{7!} + \dots \text{ to infinity.}$$

Also from equation (6.2) we get

$$\cos x = 1 - \frac{x^2}{2!} + \frac{x^4}{4!} - \frac{x^6}{6!} + \dots \text{ to infinity.}$$

Notice that x must be in radians in the above. These series are convergent for all values of x.

Example   Using the cosine series, evaluate $\cos 0.1$ rad.

$$\begin{aligned}
\cos 0.1 &= 1 - \frac{(0.1)^2}{2!} + \frac{(0.1)^4}{4!} - \frac{(0.1)^6}{6!} + \dots \\
&= 1 - 0.005 + 0.000\,004\,16 - 0.000\,000\,001 \\
&= 0.995\,004 \text{ to six decimal places.}
\end{aligned}$$

Exercise 6f

1   Write down the equations of the straight lines

   (a) of gradient 2, with intercept 3 on the y-axis.
   (b) of gradient $-\frac{1}{3}$, with intercept -1 on the y-axis.
   (c) parallel to the line $y = 3x$, with intercept 2 on the y-axis.
   (d) with gradient $\frac{1}{2}$, passing through the point $(1, -2)$.
   (e) perpendicular to the line $y = \frac{1}{2}x$, passing through the point $(-1, 3)$.

2   Find the equations of the tangent and normal at the point $(1, 1)$ on the curve $y = x^2 - 3x + 3$.

3   Find the slope of the tangent to the curve $y = xe^x - 2$ at the point $(0, -2)$. What is the equation of the normal at this point?

4   A parabola is represented by the equations $x = 2t^2$, $y = 4t$. Find the gradient of the tangent at any point in terms of the parameter t. Determine the equations of the tangent and the normal at the point on the curve where $t = 2$.

5   Find the equation of the tangent to the curve $4x^2 + 9y^2 = 72$ at the point $(3, 2)$.

6   A curve is represented in parametric form by the equations $x = t$, $y = \frac{1}{t}$.
   Find the equations of the tangent and normal at the point where $t = \frac{1}{3}$.

7   Given that $x = 2$ is an approximate root of $x^3 = 4x + 2$, use Newton's method once to find a closer value.

117   The sine and cosine series

8 Given $x = 2$ as an approximate value of a root of $x^3 - 6x + 1 = 0$, use Newton's method twice to find a closer approximation.

9 Show that there is a root of $x^3 - x + 1 = 0$ between -3 and $+2$ and find its value correct to three significant figures.

10 Find the values of $y = e^{-x} - x$ for values of x at $0, 0.2, 0.4, 0.6, 0.8, 1.0$ and draw the graph between $x = 0$ and $x = 1$.

 Show that the equation $e^{-x} - x = 0$ has a root between $x = 0.4$ and $x = 0.6$ and use Newton's method to find the root correct to three decimal places.

11 Write down the first four terms of the expansions in ascending powers of x of:

 (a) $\sin 2x$ $\qquad$ (b) $\cos ax$ $\qquad$ (c) $\cos \dfrac{x}{2}$

12 Write down the series for e^x and $\sin x$ and hence expand $e^x \sin x$ in a series of ascending powers of x as far as the term in x^3.

13 Give the first four terms in the expansion of $\sin x$ and hence find the value of $\sin 5.73°$ correct to six decimal places.

14 Expand $\log (1+x) \cos x$ as a series in ascending powers of x as far as the term in x^4.

6g Curve sketching

To know how a function will vary, it is useful to know the shape of its graph. The following rules are a help in curve sketching.

1 Examine the equation for symmetry. That is, see if the equation is unchanged when

 (a) $-x$ is put for x (curve symmetrical about y-axis).
 (b) $-y$ is put for y (curve symmetrical about x-axis).
 (c) $-x$ is put for x and $-y$ is put for y (curve symmetrical about the origin, i.e. skew symmetrical).

2 Find the sign and magnitude of y when x is very large and positive. That is, when x approaches infinity. (We write this as '$x \to +\infty$'.)
 Find the sign and magnitude of y when $x \to -\infty$.
 Find the sign and magnitude of x when $y \to +\infty$ and $y \to -\infty$.

3 Find the points where the curve cuts the axes if possible.

 (a) Put $y = 0$ and solve for x for intersection with x-axis.
 (b) Put $x = 0$ and solve for y for intersection with y-axis.

4   Find the turning points of the function where $\dfrac{dy}{dx} = 0$ and distinguish between maximum and minimum values.

5   Find the simple asymptotes if any. If $y = \dfrac{f_1(x)}{f_2(x)}$ equate the factors of the denominator to zero for simple asymptotes which are parallel to the y-axis.

   Below are sketched examples of curves of types of functions that occur in connexion with common mathematical problems.

Examples of symmetry

Figure
17

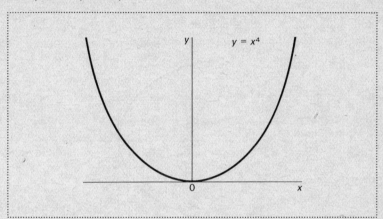

Figure
18

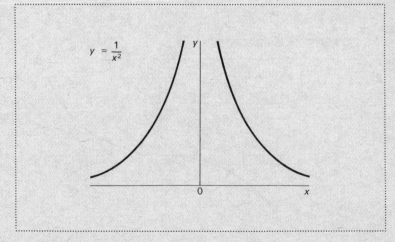

Figure
19

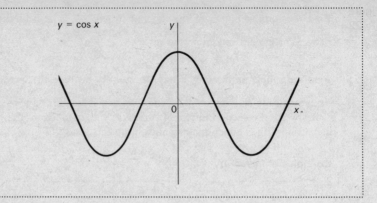

Figure
20

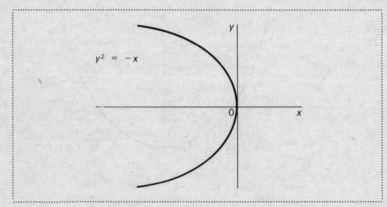

Examples of skew symmetry

Figure
21

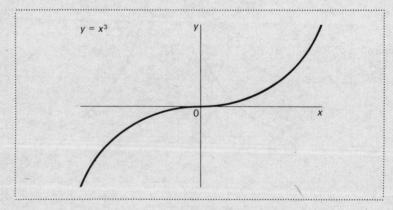

Figure
22

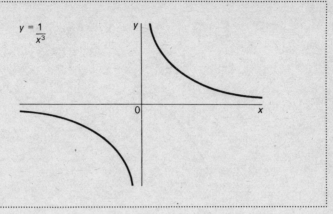

$y = \dfrac{1}{x^3}$

Figure
23

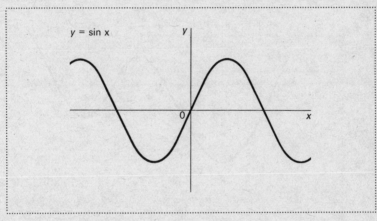

$y = \sin x$

Some curves of the form $y = f(x)$

Figure
24

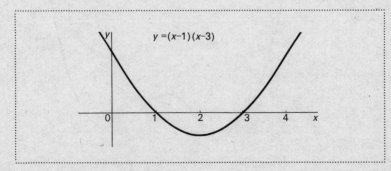

$y = (x-1)(x-3)$

Figure
25

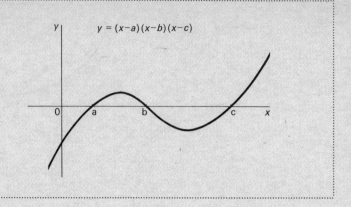

$y = (x-a)(x-b)(x-c)$

Figure
26

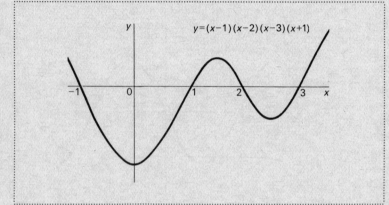

$y=(x-1)(x-2)(x-3)(x+1)$

Figure
27

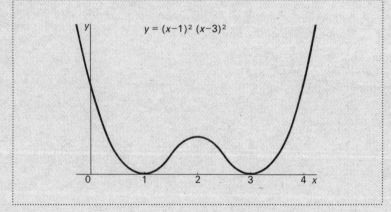

$y = (x-1)^2 (x-3)^2$

Figure
28

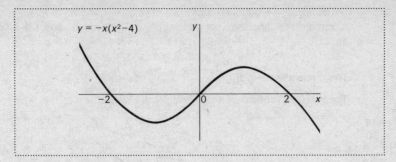

$y = -x(x^2-4)$

Curve with simple asymptotes

$$y = \frac{1}{(x-1)(x-3)}$$

When x approaches infinity, y is the reciprocal of a large number and therefore tends to zero and it is positive. We write this as

When
$$x \to +\infty \qquad y \to 0 \quad \text{and is positive.}$$
$$x \to -\infty \qquad y \to 0 \quad \text{and is still positive.}$$

Thus the curve gets closer and closer to the x-axis for large values of x both positive and negative.

The x-axis is an asymptote.

When x is slightly <1, y is positive and large.

When x is slightly >1, y is negative and large.

Figure
29

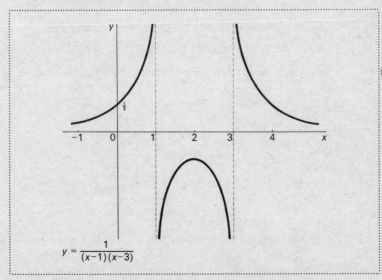

$$y = \frac{1}{(x-1)(x-3)}$$

123 Curve sketching

The line $x = 1$ is an asymptote.

Similarly the line $x = 3$ is an asymptote. The curve is shown sketched in fig. 29.

The circle $x^2 + y^2 = a^2$

The curve is symmetrical about both x- and y-axes.

Figure
30

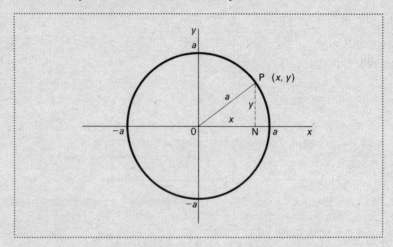

When $x = 0$, $y^2 = a^2$, i.e. $y = \pm a$

When $y = 0$, $x^2 = a^2$, i.e. $x = \pm a$

Any point P whose coordinates are x and y is at a distance

$\sqrt{(x^2 + y^2)} = \sqrt{a^2} = a$ from the origin 0.

The curve is obviously a circle centre the origin and radius a. See fig. 30.

Exercise 6g

Sketch the graphs of

1	$y = 4x^2$	2 $\quad y = -x^2$	3 $\quad x = y^2$
4	$y^2 = -16x$	5 $\quad y = -x^3$	6 $\quad xy = 4$
7	$y = 3x^5$		8 $\quad y = 1 - x^2$
9	$y = (x+1)(x-1)(x-3)$		10 $\quad y = -x(x-1)(x-3)$
11	$y = x^3 - x^2 - 2x$		12 $\quad y = (x-1)^2(2-x)$

13    $y = x(x-1)(x^2-4)$ 14 $y = (x^2-2)(4-x^2)$

15    $y = (x-1)^2(x+1)(x+3)$ 16 $y = 4x^2(x-1)^2$

17    $y = \dfrac{1}{(x+1)(x-2)}$ 18 $y = \dfrac{1}{4-x^2}$

19    $x^2+y^2 = 4$ 20 $\dfrac{x^2}{9}+\dfrac{y^2}{4} = 1$

6h   Partial differentiation

In many practical problems functions depend upon more than one variable. For example, the volume of a right circular cone of height h and radius r is given by

$$V = \tfrac{1}{3}\pi r^2 h$$

Suppose we require the rate of change of volume V with respect to the radius r, while the height h remains constant. We differentiate V with respect to r treating h as another constant like π.

$$\left[\frac{dV}{dr}\right]_{h \text{ constant}} = \tfrac{1}{3}(\pi h)\, 2r = \tfrac{2}{3}\pi rh$$

Similarly we may find the rate of change of volume V with respect to height h when r remains constant.

Thus $\left[\dfrac{dV}{dh}\right]_{r \text{ constant}} = \tfrac{1}{3}(\pi r^2)\times 1 = \tfrac{1}{3}\pi r^2$

This process of treating one variable as a constant while differentiating with respect to the other is known as *partial differentiation* and a special notation is used.

Thus $\left[\dfrac{dV}{dr}\right]_{h \text{ constant}}$ is written $\dfrac{\partial V}{\partial r}$ and is known as the partial derivative of V

with respect to r (and is read as 'partial dee V by dee r').

Similarly $\left[\dfrac{dV}{dh}\right]_{r \text{ constant}}$ is written $\dfrac{\partial V}{\partial h}$ and is the partial derivative of V with

respect to h.

Suppose V is a function of three independent variables x, y, z, and we want the partial derivative of V with respect to any variable, we differentiate in the usual way with respect to the stated variable regarding all the others as constants.

Example 1 Find $\dfrac{\partial V}{\partial x}, \dfrac{\partial V}{\partial y}, \dfrac{\partial V}{\partial z}$ if

i $V = x^2 y^3 z^4$ ii $V = xy^2 + y^2 z^3 + 2x - 3y + 4z$

i $\dfrac{\partial V}{\partial x} = 2xy^3 z^4$ ii $\dfrac{\partial V}{\partial x} = y^2 + 2$

$\dfrac{\partial V}{\partial y} = 3y^2 x^2 z^4$ $\dfrac{\partial V}{\partial y} = 2yx + 2yz^3 - 3$

$\dfrac{\partial V}{\partial z} = 4z^3 x^2 y^3$ $\dfrac{\partial V}{\partial z} = 3z^2 y^2 + 4$

As in ordinary differentiating, a partial derivative may be differentiated partially again to give second and higher order partial derivatives.

Thus if $V = x^2 y^3$, $\dfrac{\partial V}{\partial x} = 2xy^3$, $\dfrac{\partial V}{\partial y} = 3x^2 y^2$

$$\frac{\partial}{\partial x}\left(\frac{\partial V}{\partial x}\right) = \frac{\partial^2 V}{\partial x^2} = 2y^3$$

and $\dfrac{\partial}{\partial y}\left(\dfrac{\partial V}{\partial y}\right) = \dfrac{\partial^2 V}{\partial y^2} = 6x^2 y$

The total change in a function when the variables all vary

So far we have only considered cases where the variables vary one at a time.

Consider the change in the volume of the right circular cone whose volume $V = \frac{1}{3}\pi r^2 h$, when both r and h change.

Let r increase by δr and h by δh and let δV be the corresponding increase in V.

Then $V + \delta V = \dfrac{\pi}{3}(r + \delta r)^2 (h + \delta h)$

and $\delta V = \dfrac{\pi}{3}\left[r^2 h + 2rh\delta r + r^2 \delta h + h(\delta r)^2 + 2r\delta r \delta h + \delta h(\delta r)^2 - r^2 h\right]$

$\qquad = \dfrac{2\pi rh}{3}\delta r + \dfrac{\pi r^2}{3}\delta h + $ terms involving squares and higher powers of δr and δh

$\qquad = \dfrac{\partial V}{\partial r}\delta r + \dfrac{\partial V}{\partial h}\delta h + \dots$

Thus for small changes we have the approximate formula

$$\delta V = \frac{\partial V}{\partial r}\delta r + \frac{\partial V}{\partial h}\delta h$$

This result enables one to find the approximate total change in a function when all the variables alter.

If V is a function of x and y, i.e. $V = f(xy)$,

then $\quad \delta V \simeq \dfrac{\partial V}{\partial x} \delta x + \dfrac{\partial V}{\partial y} \delta y$

Similarly if V is a function of three variables x, y, z,

then $\quad \delta V \simeq \dfrac{\partial V}{\partial x} \delta x + \dfrac{\partial V}{\partial y} \delta y + \dfrac{\partial V}{\partial z} \delta z$

Example 1 Find the approximate change in the volume V of a cone if the radius r increases from 6 cm to 6·05 cm and the height decreases from 8 cm to 7·96 cm.

Since V is a function of r and h

$$\delta V \simeq \frac{\partial V}{\partial r} \delta r + \frac{\partial V}{\partial h} \delta h$$

$$V = \tfrac{1}{3}\pi r^2 h \qquad \frac{\partial V}{\partial r} = \frac{2\pi r h}{3}; \frac{\partial V}{\partial h} = \frac{\pi r^2}{3}$$

$$r = 6, \delta r = 0\cdot05, h = 8, \delta h = -0\cdot04$$

$$\delta V \simeq \frac{2\pi \times 6 \times 8}{3}(0\cdot05) + \frac{\pi \times 6^2}{3}(-0\cdot04)$$

$$= 1\cdot6\pi - 0\cdot48\pi = 1\cdot12\pi \text{ cm}^3$$
$$= 3\cdot52 \text{ cm}^3$$

Example 2 The circumferential or hoop stress p in a cylindrical shell of mean radius r and thickness t when subjected to an internal pressure P is given by $p = \dfrac{Pr}{t}$. This formula is used to calculate the hoop stress in a cylindrical boiler estimating the diameter to be 1·44 m, the plate thickness 0·5 cm and the maximum working pressure to be 300 kN m^{-2}. What is the approximate error in the calculated stress if the actual values are diameter 1·50 m, plate thickness 0·47 cm, and internal pressure 320 kN m^{-2}?

Since p is a function of P, r and t, then

$$\delta p \simeq \frac{\partial p}{\partial P} \delta P + \frac{\partial p}{\partial r} \times \delta r + \frac{\partial p}{\partial t} \times \delta t$$

$$p = \frac{Pr}{t} \qquad \frac{\partial p}{\partial P} = \frac{r}{t}; \frac{\partial p}{\partial r} = \frac{P}{t}; \frac{\partial p}{\partial t} = -\frac{Pr}{t^2}$$

$$P = 300 \text{ kN m}^{-2}, r = 0\cdot72 \text{ m}, t = 0\cdot005 \text{ m}$$
$$\delta P = 20 \text{ kN m}^{-2}, \delta r = 0\cdot03 \text{ m}, \delta t = -0\cdot0003 \text{ m}$$

$$\delta p \simeq \frac{0 \cdot 72}{0 \cdot 005} \times 20 + \frac{300}{0 \cdot 005} \times 0 \cdot 03 - \frac{300 \times 0 \cdot 72}{(0 \cdot 005)} (-0 \cdot 0003)$$

$$\simeq 2880 + 1800 + 2592$$

$$= 7272 \text{ kN m}^{-2} \text{ approximately.}$$

Thus the actual stress is approximately $7 \cdot 3 \text{ MN m}^{-2}$ above the estimated value.

Exercise 6h

In examples 1–6 find $\dfrac{\partial z}{\partial x}$ and $\dfrac{\partial z}{\partial y}$.

1 $z = xy$

2 $z = x^2 y^3 + 2x + y$

3 $z = \dfrac{3x}{y}$

4 $z = x^3 y + \dfrac{y}{x^2} - \dfrac{1}{y}$

5 $z = \sin(3x + 2y)$

6 $z = \sin 2x \cos 3y$

7 If $pv = RT$, find $\dfrac{\partial p}{\partial v}$ and $\dfrac{\partial p}{\partial T}$.

8 If $p = \dfrac{Pr}{t}$, find $\dfrac{\partial p}{\partial P}, \dfrac{\partial p}{\partial r}, \dfrac{\partial p}{\partial t}$.

9 If $T = 2\pi \sqrt{\dfrac{l}{g}}$, find $\dfrac{\partial T}{\partial l}$ and $\dfrac{\partial T}{\partial g}$.

10 If $h = \dfrac{flQ^2}{10d^5}$, find $\dfrac{\partial h}{\partial l}, \dfrac{\partial h}{\partial Q}, \dfrac{\partial h}{\partial d}$.

11 If $p = \dfrac{Pr}{t}$, show that $\delta p \simeq \dfrac{Pr}{t} \left(\dfrac{\delta P}{P} + \dfrac{\delta r}{r} - \dfrac{\delta t}{t} \right)$.

12 The stress p due to rotation in the thin rim of a fly-wheel is given by $p = \dfrac{Wr^2\omega^2}{g}$ where r is the mean radius, ω the angular velocity and W the specific weight. If r and ω increase by δr and $\delta \omega$ respectively, find an expression for δp and show that $\dfrac{\delta p}{p} = 2 \left(\dfrac{\delta r}{r} + \dfrac{\delta \omega}{\omega} \right)$ approximately.

If the radius increases by 2% and the speed by $1\frac{1}{2}\%$ show that the stress increases by 7%.

13 Find the approximate error in calculating the volume of a right circular cylinder of height 6·5 cm and radius of base 4·4 cm if an error of $+0·03$ cm is made in the height and $-0·05$ cm in the radius.

14 Find the approximate change in volume of a rectangular block of length 20 cm, breadth 15 cm and depth 12 cm if the length increases by 1 cm and the breadth and depth each decrease by $\frac{1}{2}$ cm.

15 The current i in a circuit is given by $i = \dfrac{V}{R}$.

Find the approximate change in the current if V falls from 240 to 230 volts and R increases from 50 to 52 ohms.

16 The flow of water over a rectangular notch is given by $Q = 1·84\, LH^{\frac{3}{2}}\ \mathrm{m^3\ s^{-1}}$ where L is the width of the notch in m and H the depth of water in m.

Find a formula for the change in flow δQ due to small changes δL in L and δH in H.

Calculate the approximate change in flow over a rectangular notch 2 m wide if the depth of water changes from 30 cm to 28 cm when the width of the notch is increased by 5 cm.

Miscellaneous exercises 6

1 (a) A closed cylindrical can is sealed by soldering up one side and around the top and bottom edges. The solder is applied uniformly and the total volume enclosed is 10 $\mathrm{cm^3}$. Find the ratio of the radius of the can to its height in order that the minimum amount of solder is used.

  (b) Sketch (do not plot on squared paper) the curve $y = \log_e x$. Find the value of x for which the tangent to this curve is parallel to the line $y = 2x$.

2 (a) Find the maximum and minimum values of the function $y = \dfrac{(x-1)(x-6)}{x-10}$

  and give a sketch of the function.

  (b) A wire of length L is bent so as to form the boundary of a sector of a circle of radius r and angle θ radians. Show that $\theta = \dfrac{L-2r}{r}$.

  Prove that the area of the sector is greatest when the radius is $\dfrac{L}{4}$.

3 (a) The distance x m which a body travels in a time t s is given by

  $x = 5+6t+4t^2$

  Find the velocity and acceleration of the body after 3 s.

(b) A body is x m from a point O on a horizontal track and is pulled along the track by a rope which passes over a pulley 5 m vertically above O. Show that if the rope is hauled in at a rate $\dfrac{dl}{dt}$ then $x\dfrac{dx}{dt} = l\dfrac{dl}{dt}$ where l is the length of rope between the body and the pulley.

If the rope is hauled in at 2 m s^{-1} find the speed of the body when $x = 12$ m.

4   (a) Find the maximum and minimum values of $y = \dfrac{2x+1}{x(x-4)}$. Sketch the graph of the function.

(b) At a certain port t hours after high water, the height h m of the tide above a fixed datum level is given by $h = 3 + 3\cos\dfrac{\pi t}{6}$ approximately.

At what rate in millimetres per second is the tide falling two hours after high water?

5   Find the maximum and minimum values of the curve given by

$6(y - x^4) = 8x^3 - 9x^2 + 11$

Show them on a sketch.   *U.L.C.I.*

6   (a) Determine the maximum value of the function

$$M = c^2 - \frac{(a+b)}{2}x - w^2x^2$$

and the value of x at which this maximum occurs, expressing the results in terms of the constants a, b, c and w.

If $a = 5\cdot5, b = 9\cdot5, c = 15, w = \sqrt{3}$; evaluate the maximum value of M and the corresponding value of x.

(b) A curve has the equation $y = \dfrac{x}{4} + \dfrac{4}{x}$. Show that y has a maximum value of -2. Find its minimum value and sketch the curve.

                                                                                U.E.I.

7   (a) Find the acute angle θ for which the function $\dfrac{2 - \sin\theta}{\cos\theta}$ is a minimum and sketch a graph of the function for $0 \leqslant \theta \leqslant \pi/2$.

(b) A cylindrical hole is bored centrally through a hemispherical casting of radius a perpendicular to its flat face. Taking x as the length of the cylinder wall, express the volume of the hole as a function of x.

Determine the ratio of the length of the hole to its radius when the volume of the hole is a maximum.

                                                                                U.E.I.

8   The cross-section of an open irrigation canal is an isosceles trapezium. The bottom, which is the shorter of the two parallel sides, is of length x; the sloping

sides, each of length y, are inclined to the horizontal at an angle α where $\tan \alpha = \frac{4}{3}$. If the area of the cross-section is 448, show that $(5x+3y)y = 2800$, and deduce that the area of the retaining surface (bottom + sloping sides) is given by

$$A = L\left[\frac{560}{y}+\frac{7}{5}y\right]$$

where L is the length of the canal.

Find the dimensions of the cross-sectional area if the retaining surface is to be a minimum, giving a test for the nature of the extreme value.

E.M.E.U.

9 (a) The function $y = 2x^3+ax^2+bx+3$, where a and b are constants, has turning points at $x = 1$ and $x = -2$.

Find the values of a and b. Calculate the values of y at the two turning points and state which is a maximum and which is a minimum. Find also the point of inflexion.

(b) The current I in a circuit is given by $I = \dfrac{E}{R+r}$ where $E = 10$ volts, $R = 20$ ohms and both are constant.

Find the rate of change of current when $r = 10$ ohms and is increasing at the rate of 3 ohms per minute.

D.T.C.

10 (a) Calculate the gradients of the curve $y = e^{x^2}$ at the points where $y = 2$.

(b) A manufacturer can sell x items per week at a selling price of £$(1000-\frac{1}{2}x)$ per item, and the total production cost of the x items is £$(450x+11000)$. Calculate his maximum weekly profit.

If now the government imposes a tax amounting to 10% of the selling price of each item sold, find the reduction in the maximum weekly profit and how many fewer items must be sold to achieve the maximum.

E.M.E.U.

11 Write down the equation of the straight line of gradient m which passes through the point (x_1, y_1).

Find the equation of the tangent to the circle $x^2+y^2 = 5$ at the point $(1, 2)$.

12 (a) If $z = \sec x$ find the approximate change in z if x changes from $56° 30'$ to $56° 34'$.

(b) If $z = 2x^2y^3$ find the approximate change in z if x changes from 10 to 10·05 and y changes from 6 to 5·97.

13 (a) If $z = \cos(2x-3y)$ find $\dfrac{\partial z}{\partial x}$ and $\dfrac{\partial z}{\partial y}$.

(b) If $p = \dfrac{200r}{t}$ find $\dfrac{\partial p}{\partial r}$ and $\dfrac{\partial p}{\partial t}$ and show that for small increments δr in r and δt in t the change δp in p is given by

$$\frac{\delta p}{p} = \frac{\delta r}{r} - \frac{\delta t}{t}$$

14　Find the values of $y = e^{-x} - 2x$ for values of x at 0, 0·2, 0·4, 0·6, and draw the graph of the function between $x = 0$ and $x = 0\cdot6$.

Show that the equation $e^{-x} - 2x = 0$ has a root at approximately 0·4 and taking this as a first approximation, use Newton's method to find the root correct to two decimal places.

15　Given that $x = 1\cdot5$ is an approximate root of $x^3 + \dfrac{3x}{2} - 2 = 0$, use Newton's method twice to find a closer approximation.

16　Show that the equation $x + e^x = 0$ has a root between -1 and 0 and find its value correct to two decimal places.

17　The equation $\tan \theta + 0\cdot8\theta = 0$ arises in a problem on a framework. Sketch the graphs of $y = \tan \theta$ and $y = -0\cdot8\theta$ where θ is in radians from 0 to π radians and show that a solution of $\tan \theta + 0\cdot8\theta$ is approximately 2 radians. Use Newton's method once to find a closer value.

18　The circumferential or hoop stress p in a cylindrical shell of mean radius r and thickness t subjected to an internal pressure P is given by $p = \dfrac{Pr}{t}$.

Calculate the hoop stress expected in a cylindrical boiler designed to have a diameter 1·2 m, plate thickness 0·5 cm and internal pressure 400 kN m^{-2}.

If the boiler, when manufactured, has a diameter 1·25 m, a plate thickness 0·46 cm, and is subjected to an internal pressure of 415 kN m^{-2}, use partial differentiation to find approximately the change in hoop stress.

19　A helical spring, when manufactured, differs from the assumed design data as follows; wire diameter d 3% too great, coil radius R 2% too small.

The maximum shear stress q in the spring is given by $q = \dfrac{16WR}{\pi d^3}$ where W is the load. By partial differentiation find the approximate change δq in q in terms of small changes δR and δd in R and d. Hence find approximately the percentage change in the maximum shear stress for the spring.

20　The deflexion y at the centre of a rod, diameter d, length l, supported horizontally at its ends and carrying a load W at its mid point is given by

$y = \dfrac{kWl^3}{d^4}$ where k is a constant depending on the material.

Show that the change in deflection δy due to small changes δW in W, δl in l, and δd in d, is given by

$$\delta y \simeq \frac{kWl^3}{d^4}\left(\frac{\delta W}{W}+\frac{3\delta l}{l}-\frac{4\delta d}{d}\right)$$

Hence find the percentage change in the deflexion at the centre if the load W increases by 3%, the span l decreases by 2% and the diameter d increases by 1%.

7 Trigonometric equations

7a General solution of trigonometric equations

In volume 1, section 11d we showed that trigonometric equations differ from algebraic in that they often have an unlimited number of solutions.

For example solutions of the equation $\sin x = \frac{1}{2}$ are given by any angle in the first or second quadrant which makes an angle of $30°$ with the horizontal line, i.e. angles xOA or xOB in fig. 31.

Figure 31

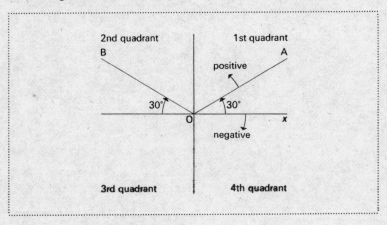

The following are all solutions.

$30°, 150°, (360+30)°, (360+150)°, (720+30)°, -210°$, etc.

These can all be represented by the general formulae

$(360n+30)°$ and $(360n+150)°$ where n takes the values 0, 1, 2, 3, etc. and -1, -2, -3, etc.

Similarly solutions of $\cos x = \frac{1}{2}$ are angles

$60°, -60°, (360-60)° = 300°, (360+60)°, (720-60)°$, etc. (See fig. 32.)

There are generally two angles between $0°$ and $360°$ corresponding to a numerical value of a trigonometric ratio.

Figure 32

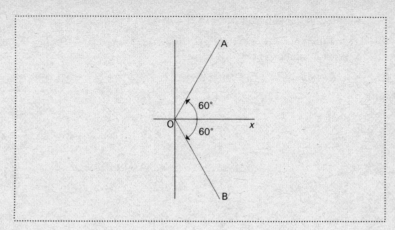

The solution of $\sin x = \sin \alpha^\circ$

Solutions are $x = \alpha^\circ, (180-\alpha)^\circ, (360+\alpha)^\circ, (540-\alpha)^\circ$, etc. These are all contained in the formulae

$x = (360n+\alpha)^\circ$ and $(360n+180-\alpha)^\circ$
$\quad = (2n \times 180+\alpha)^\circ$ and $[(2n+1)180-\alpha]^\circ$

where $n = 0, 1, 2, 3, \ldots$

A single formula for these solutions is

$x = [180n+(-1)^n\alpha]^\circ$

If n is an odd whole number then the term $(-1)^n$ is -1 whereas if n is an even number including zero $(-1)^n$ is $+1$. This solution is known as the *general solution*.

Example 1 Find the general solution of the equation

$\sin x = 0.6324$

From the tables we see that 0.6324 is the sine of the acute angle $39^\circ\ 14'$.

General solution is $x = 180n^\circ +(-1)^n\ 39^\circ\ 14'$

where n takes values 0, 1, 2 etc.; note $(-1)^0 = 1$.

The general solution of $\cos x = \cos \alpha^\circ$

In this case solutions are $x = \alpha^\circ, (360-\alpha)^\circ, (360+\alpha)^\circ, (720-\alpha)^\circ$, etc.

or $x = (360n \pm \alpha)^\circ$ $n = 0, 1, 2$, etc.

The general solution of $\tan x = \tan \alpha°$

Solutions are $x = \alpha°, (180+\alpha)°, (360+\alpha)°, (540+\alpha)°$, etc. (see fig. 33) and the general solution is

$$x = (180n+\alpha)° \qquad n = 0, 1, 2, \text{etc.}$$

Figure 33

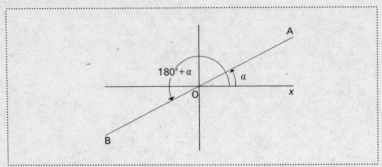

In solving trigonometric equations try to reduce the given equation to one of the standard equations given below, where α is the numerical value of the smallest positive angle satisfying the equation.

Equation	General solution
$\sin x = \sin \alpha$	$x = [180n+(-1)^n\alpha]°$ or if α is in radians $x = [n\pi+(-1)^n\alpha]^c$
$\cos x = \cos \alpha$	$x = (360n\pm\alpha)°$ or in radians $x = (2n\pi\pm\alpha)^c$
$\tan x = \tan \alpha$	$x = (180n+\alpha)°$ or in radians $x = (n\pi+\alpha)^c$

Example 2 Find the general solution of the equation

$2 \sin 2x - \cos 2x = 0$ and give all the solutions between $0°$ and $360°$.

$2 \sin 2x = \cos 2x$

Dividing both sides by $2 \cos 2x$ gives

$\tan 2x = \tfrac{1}{2} = 0\cdot5$

From the tables $\tan 26° \, 34' = 0\cdot5$

$\tan 2x = \tan 26° \, 34'$
$\quad 2x = 180n° + 26° \, 34'$
and $x = 90n° + 13°17'$ is the general solution.

For numerical values put $n = 0, 1, 2,$ etc. in turn. Numerical solutions between $0°$ and $360°$ are

$13°\ 17',\ 103°\ 17',\ 193°\ 17',\ 283°\ 17'$

Example 3 Find the general solutions of the equation

$2 \sin \theta\ (3 \cos 2\theta + 2) = \sin 2\theta$

Put all the terms on the left hand side and substitute $2 \sin \theta \cos \theta$ for $\sin 2\theta$ and $2 \cos^2 \theta - 1$ for $\cos 2\theta$. The equation becomes

$2 \sin \theta\ (6 \cos^2 \theta - 3 + 2) - 2 \sin \theta \cos \theta = 0$
i.e. $\qquad 2 \sin \theta\ (6 \cos^2 \theta - 1 - \cos \theta) = 0$

This factorizes still further into

$\qquad 2 \sin \theta\ (2 \cos \theta - 1)(3 \cos \theta + 1) = 0$

Thus $\quad \sin \theta = 0 \qquad 2 \cos \theta - 1 = 0 \quad$ or $\quad 3 \cos \theta + 1 = 0$
i.e. $\quad \sin \theta = 0 \qquad \cos \theta \quad = \frac{1}{2} \quad$ or $\quad \cos \theta \quad = -\frac{1}{3}$
i $\ \sin \theta = 0 \quad$ gives $\quad \theta = n\pi$ radians

ii $\ \cos \theta = \dfrac{1}{2} = \cos 60° = \cos \dfrac{\pi}{3}$

$\qquad \theta = 2n\pi \pm \dfrac{\pi}{3}$

iii $\quad \cos \theta = -0 \cdot 3333 = \cos (180° - 70°\ 32')$
$\qquad\qquad\quad = \cos 109°\ 28'$
$\qquad\qquad \theta = 360n° \pm 109°\ 28'$
or $\qquad = (2n\pi \pm 1 \cdot 91)$

Thus the general solutions are $n\pi,\ 2n\pi \pm \dfrac{\pi}{3},\ 2n\pi \pm 1 \cdot 91.$

Example 4 Solve $\cos 7\theta + \cos 3\theta = \cos 5\theta + \cos \theta.$

Using the formula $\quad \cos + \cos = 2 \cos (\frac{1}{2}$ sum$) \cos (\frac{1}{2}$ difference$)$
gives $\qquad\qquad 2 \cos 5\theta \cos 2\theta = 2 \cos 3\theta \cos 2\theta$

Putting all the terms on one side and factorizing gives

$2 \cos 2\theta\ (\cos 5\theta - \cos 3\theta) = 0$

Hence $\quad \cos 2\theta = 0 = \cos \dfrac{\pi}{2}$

and $\qquad 2\theta = 2n\pi \pm \dfrac{\pi}{2}$

whence $\qquad \theta = n\pi \pm \dfrac{\pi}{4}$

or $\quad \cos 5\theta - \cos 3\theta = 0$

i.e. $\qquad \cos 5\theta = \cos 3\theta$

$$5\theta = 2n\pi \pm 3\theta$$

Taking the + sign

$$5\theta = 2n\pi + 3\theta$$
$$2\theta = 2n\pi$$
$$\theta = n\pi$$

Taking the − sign

$$5\theta = 2n\pi - 3\theta$$
$$8\theta = 2n\pi$$

$$\theta = \frac{n\pi}{4}$$

The solutions are $n\pi \pm \dfrac{\pi}{4}, \; n\pi, \; \dfrac{n\pi}{4}$.

Notice that these are all contained in the solution

$$\theta = \frac{n\pi}{4}$$

Exercise 7a

Obtain the general solutions of the following equations and in questions 1 to 10 give all solutions between $0°$ and $360°$.

1   $\sin \theta = \dfrac{1}{\sqrt{2}}$ 2 $\sin^2 \theta = \dfrac{1}{4}$ 3 $\tan \theta = \sqrt{3}$

4   $\cos \theta = -\dfrac{1}{2}$ 5 $\tan \theta = -1$ 6 $\cot \theta = 1$

7   $\sec \theta = 2$ 8 $\operatorname{cosec} \theta = \dfrac{2}{\sqrt{3}}$

9   What is the most general value of θ that satisfies both of the equations:

     (a) $\cos \theta = -\dfrac{1}{\sqrt{2}}$ and $\tan \theta = 1$

     (b) $\sin \theta = -\dfrac{1}{2}$ and $\tan \theta = \dfrac{1}{\sqrt{3}}$

10   $\cos^2 \theta - \sin \theta - \dfrac{1}{4} = 0$

11 $2 \sin^2 \theta + 3 \cos \theta = 0$ 12 $2\sqrt{3} \cos^2 \theta = \sin \theta$

13 $\cos \theta + \cos^2 \theta = 1$ 14 $4 \cos \theta - 3 \sec \theta = 2 \tan \theta$

15 $\tan^2 \theta - (1 + \sqrt{3}) \tan \theta + \sqrt{3} = 0$ 16 $\sin 5\theta = \dfrac{1}{\sqrt{2}}$

17 $\cos 3\theta = 0$ 18 $\sin 9\theta = \sin \theta$

19 $\sin 3\theta = \sin 2\theta$ 20 $\cos 5\theta = \cos 4\theta$

21 $\tan 3\theta = \cot \theta$ 22 $\sin \theta + \sin 7\theta = \sin 4\theta$

23 $\cos \theta + \cos 7\theta = \cos 4\theta$ 24 $\cos \theta + \cos 3\theta = 2 \cos 2\theta$

25 $\sin \theta + \sin 3\theta + \sin 5\theta = 0$

7b To solve an equation of the type $a \cos \theta + b \sin \theta = c$

There are two main methods.

Method 1   Using the identity $a \cos \theta + b \sin \theta = \sqrt{(a^2 + b^2)} \cos (\theta - \alpha)$ where $\tan \alpha = \dfrac{b}{a}$.

Example   Solve $2 \cos \theta + 3 \sin \theta = 2 \cdot 5$.

The left hand side may be written $\sqrt{(2^2 + 3^2)} \cos (\theta - \alpha)$ where $\tan \alpha = \dfrac{3}{2}$,

i.e. $\tan \alpha = 1 \cdot 5 = \tan 56° \ 19'$

The equation becomes

$\sqrt{13} \cos (\theta - 56° \ 19') = 2 \cdot 5$

$\qquad \cos (\theta - 56° \ 19') = \dfrac{2 \cdot 5}{\sqrt{13}} = 0 \cdot 693$

$\qquad \qquad \qquad \qquad \quad = \cos 46° \ 6'$

$\qquad \theta - 56° \ 19' = 360n° \pm 46° \ 6'$

$\theta = 360n° + 46° \ 6' + 56° \ 19' \text{ or } 360n° - 46° \ 6' + 56° \ 19'$

$\quad = 360n° + 102° \ 25' \qquad \text{or } 360n° + 10° \ 13'$

It is generally better to use the form $\sqrt{(a^2 + b^2)} \cos (\theta - \alpha)$ rather than $\sqrt{(a^2 + b^2)} \sin (\theta + \beta)$, as the general solution of $\cos \theta = a$ is more compact than $\sin \theta = b$.

139   To solve an equation of the type $a \cos \theta + b \sin \theta = c$

Method 2 Using the identities $\sin \theta = \dfrac{2t}{1+t^2}$, $\cos \theta = \dfrac{1-t^2}{1+t^2}$,

where $t = \tan \dfrac{\theta}{2}$

These identities can be proved by expanding $\sin \theta$ and $\cos \theta$ in terms of the half angle $\dfrac{\theta}{2}$.

$$\sin \theta = 2 \sin \frac{\theta}{2} \cos \frac{\theta}{2} \quad \text{and} \quad \cos \theta = \cos^2 \frac{\theta}{2} - \sin^2 \frac{\theta}{2}$$

Dividing by $\cos^2 \dfrac{\theta}{2} + \sin^2 \dfrac{\theta}{2}$ which equals 1 we have

$$\sin \theta = \frac{2 \sin \frac{\theta}{2} \cos \frac{\theta}{2}}{\cos^2 \theta + \sin^2 \frac{\theta}{2}} \quad \text{and} \quad \cos \theta = \frac{\cos^2 \frac{\theta}{2} - \sin^2 \frac{\theta}{2}}{\cos^2 \frac{\theta}{2} + \sin^2 \frac{\theta}{2}}$$

Dividing numerator and denominator by $\cos^2 \dfrac{\theta}{2}$ gives

$$\sin \theta = \frac{2 \tan \frac{\theta}{2}}{1 + \tan^2 \frac{\theta}{2}} = \frac{2t}{1+t^2} \quad \text{and} \quad \cos \theta = \frac{1 - \tan^2 \frac{\theta}{2}}{1 + \tan^2 \frac{\theta}{2}} = \frac{1-t^2}{1+t^2}$$

Example Solve $2 \cos \theta + 3 \sin \theta = 2 \cdot 5$

Substituting for $\cos \theta$ and $\sin \theta$ the equation becomes

$$2 \left(\frac{1-t^2}{1+t^2} \right) + 3 \left(\frac{2t}{1+t^2} \right) = 2 \cdot 5$$

$$2 - 2t^2 + 6t = 2 \cdot 5(1 + t^2)$$
$$4 \cdot 5 t^2 - 6t + 0 \cdot 5 = 0$$
i.e. $\quad 9t^2 - 12t + 1 = 0$

This is a quadratic equation in t which has roots

$$t = \frac{12 \pm \sqrt{(12^2 - 4 \times 9)}}{2 \cdot 9} = \frac{12 \pm \sqrt{108}}{18} = 1 \cdot 244 \text{ or } 0 \cdot 0893$$

i.e. $\quad \tan \dfrac{\theta}{2} = 1 \cdot 244 = \tan 51° \ 12'$

$$\frac{\theta}{2} = 180n° + 51° \ 12'$$

or $\quad \tan \dfrac{\theta}{2} = 0.0893 = \tan 5° \, 6'$

$$\dfrac{\theta}{2} = 180n° + 5° \, 6'$$

The general solutions are $\theta = 360n° + 102° \, 24'$ and $360n° + 10° \, 12'$.

Exercise 7b

Find the general solutions of the following equations and write down all numerical solutions between $0°$ and $360°$

1 $5 \cos \theta - 2 \sin \theta = 2$

2 $5 \sin \theta + 2 \cos \theta = 5$

3 $\sqrt{3} \cos \theta + \sin \theta = \sqrt{2}$

4 $\cos \theta - \sqrt{3} \sin \theta = 1$

5 $2 \cos \theta + 3 \sin \theta = 1$

6 $15 \sin \theta + 8 \cos \theta = 10$

7 $3 \cos x + 4 \sin x = 1.5$

8 $0.18 \cos \theta + 0.65 \sin \theta = 0.66$

9 $3.2 \sin \theta - 6 \cos \theta = 1.7$

10 $\cot x + 1 - \operatorname{cosec} x = 0$

7c Inverse trigonometric (or circular) functions

If $x = \sin y$ then y is the angle whose sine is x. For the expression 'the angle whose sine is x', the abbreviation $\sin^{-1} x$ is used. Another form is $\arcsin x$. The function $\sin^{-1} x$ is known as the inverse sine of x.

The symbol $(^{-1})$ must not be confused with a power in the way that $\sin^2 x$ is used for $(\sin x)^2$.

The graph of $y = \sin^{-1} x$ is shown in fig. 34 and may be obtained from the graph of $y = \sin x$ by interchanging the axes of x and y.

The angle whose sine is $\frac{1}{2}$ may mean any of the angles $30°$, $150°$, $390°$, $-210°$, etc. In order to avoid any ambiguity as to which angle is meant, the *principal value* of $\sin^{-1} x$ is defined as the value which lies between $-\dfrac{\pi}{2}$ and $+\dfrac{\pi}{2}$. The principal value lies on the portion of the graph AB shown in thick line in fig. 34.

Unless otherwise stated, $\sin^{-1} x$ is taken to mean the principal value only.

Thus $\quad \sin^{-1} \dfrac{1}{2} = 30° = \dfrac{\pi}{6}$ rad

In the same way $\cos^{-1} x$ is used for the angle whose cosine is x, $\cot^{-1} x$ for

Figure
34

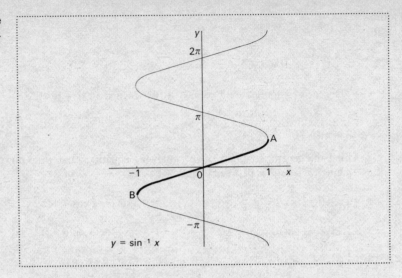

$y = \sin^{-1} x$

the angle whose cotangent is x, etc. The range for the principal values of the inverse trigonometric functions is from $-\dfrac{\pi}{2}$ to $\dfrac{\pi}{2}$ for all the functions except the inverse cosine and inverse secant. The principal values for $\cos^{-1} x$ and $\sec^{-1} x$ are restricted to the range 0 to π.

The graphs of $y = \cos^{-1} x$ and $y = \tan^{-1} x$ are shown in figs 35 and 36. the principal values being restricted to the thick-line portion AB of each graph.

Figure
35

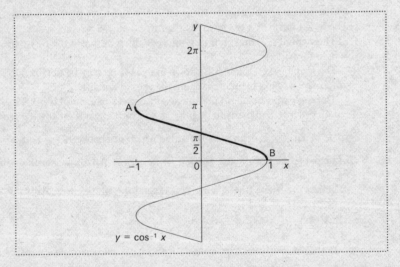

$y = \cos^{-1} x$

Figure
36

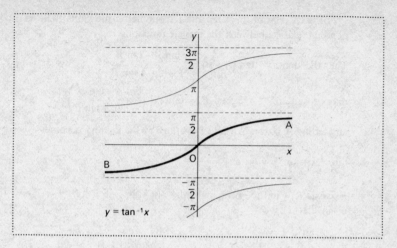

$y = \tan^{-1} x$

Relations between inverse trigonometric functions

Consider the right-angled triangle ABC in fig. 37. Letting $BC = x$ and $AC = 1$, then $AB = \sqrt{(1-x^2)}$.

Figure
37

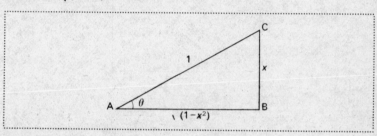

$$\theta = \sin^{-1} x = \cos^{-1} \sqrt{(1-x^2)} = \tan^{-1} \frac{x}{\sqrt{(1-x^2)}}$$

$$= \operatorname{cosec}^{-1} \frac{1}{x} = \sec^{-1} \frac{1}{\sqrt{(1-x^2)}} = \cot^{-1} \frac{\sqrt{(1-x^2)}}{x}$$

Also since $\quad \angle BAC + \angle BCA = \dfrac{\pi}{2}$

$$\sin^{-1} x + \cos^{-1} x = \frac{\pi}{2}$$

and $\quad \sin^{-1} x + \sin^{-1} \sqrt{(1-x^2)} = \dfrac{\pi}{2}$

Other relations between the inverse functions may be easily found.

For each relation between the trigonometric functions there is a corresponding relation between the inverse functions.

Thus the identity $\quad \tan(A+B) = \dfrac{\tan A + \tan B}{1 - \tan A \tan B}$

may be written $\qquad A+B = \tan^{-1}\left(\dfrac{\tan A + \tan B}{1 - \tan A \tan B}\right)$

and letting $A = \tan^{-1} x$ and $B = \tan^{-1} y$ the identity becomes

$$\tan^{-1} x + \tan^{-1} y = \tan^{-1} \dfrac{x+y}{1-xy}$$

Example 1 Find the principal values of i $\tan^{-1} 1$, ii $\cos^{-1} \frac{1}{2}$, iii $\cos^{-1}(-\frac{1}{2})$, iv $\sin^{-1}(-1)$, v $\cot^{-1} 2$.

i $\quad \tan^{-1} 1 = 45°$ 　　　　　 ii $\quad \cos^{-1} \frac{1}{2} = 60°$
iii $\cos^{-1}(-\frac{1}{2}) = 120°$ 　　　 iv $\sin^{-1}(-1) = -90°$
v $\quad \cot^{-1} 2 = 26° \, 34'$

Figure 38

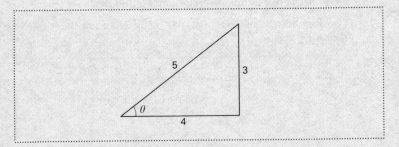

Example 2 Show that $\sin^{-1} \dfrac{3}{5} = \cos^{-1} \dfrac{4}{5} = \tan^{-1} \dfrac{3}{4}$.

From fig. 38 θ may be stated as the angle whose sin is $\dfrac{3}{5}$, whose cosine is $\dfrac{4}{5}$ and tangent $\dfrac{3}{4}$.

i.e. $\quad \theta = \sin^{-1} \dfrac{3}{5} = \cos^{-1} \dfrac{4}{5} = \tan^{-1} \dfrac{3}{4}$.

Exercise 7c

1　Find the principal values of $\sin^{-1} \dfrac{\sqrt{3}}{2}$, $\sin^{-1}\left(-\dfrac{1}{2}\right)$, $\sin^{-1} \sqrt{\tfrac{1}{2}}$, $\operatorname{cosec}^{-1} 2$,

$\operatorname{cosec}^{-1} 3$, $\sec^{-1} 1{\cdot}5$.

2 Show that $\sin^{-1}\dfrac{5}{13} = \sec^{-1}\dfrac{13}{12} = \cot^{-1}\dfrac{12}{5}$.

3 Find a value for $\sin^{-1}\dfrac{1}{2}+\cos^{-1}\dfrac{1}{3}$.

4 Prove $\sin^{-1}\dfrac{1}{\sqrt{10}}+\sin^{-1}\dfrac{3}{\sqrt{10}} = \dfrac{\pi}{2}$.

5 Find the value of $\tan^{-1}\dfrac{1}{2}+\cot^{-1}\dfrac{1}{2}$.

6 Show $\tan^{-1}\dfrac{1}{2}+\tan^{-1}\dfrac{1}{4} = \tan^{-1}\dfrac{6}{7}$.

Miscellaneous exercises 7

1 Solve the following trigonometric equations for values of x from $0°$ to $360°$.

(a) $8\sin x - 15\cos x = 4.25$

(b) $\cos 2x = 3\sin x + 2$

(c) $\dfrac{\cos 3x + \cos 2x}{\sin 3x - \sin 2x} = \dfrac{1}{\sqrt{3}}$ *U.E.I.*

2 (a) Find the values of θ between $0°$ and $360°$ which satisfy the equation $2\cos 2\theta = 7\sin \theta$.

(b) If $\tan\frac{1}{2}x = t$, show that

$$\sin x = \dfrac{2t}{1+t^2} \qquad \cos x = \dfrac{1-t^2}{1+t^2}$$

Hence or otherwise find the values of x between $0°$ and $360°$ which satisfy the equation $8\cos x - \sin x = 4$.   *D.T.C.*

3 (a) Determine the positive number R and the angle α in degrees if

$$8\cos(x-30°)+9\sin(x+60°) = R\sin(x+\alpha)$$

(b) Hence or otherwise solve the equation

$$16\cos(x-30°)+18\sin(x+60°) = -17$$

for those values of x lying between $0°$ and $360°$.   *N.C.T.E.C.*

4 (a) Prove that

$$\dfrac{\sin x - \sin 3x}{\cos x - \cos 3x} = \tan\left(\dfrac{\pi}{2}+2x\right)$$

(b) Find acute angles A, B such that

$$\cos A \cos B = 0.6 \qquad \sin A \sin B = 0.1$$

(c) Find all the values of θ between $0°$ and $360°$ for which $\cos 2\theta = 3 \cos \theta + 1$.

E.M.E.U.

5   A point P moves along a straight line OX and after t seconds its displacement x metres from the fixed point O is given by the equation

$$x = 24 \sin t - 7 \cos t$$

Express x in the form $R \sin (t - \alpha)$, evaluating the constants R and α (in radians).

Find the smallest positive value of t for which

(a) $x = 15$ (b) $x = -15$

Prove that at these instants P is moving away from O with speed 20 m s^{-1}.

6   (a) Find the positive values of θ ($< 90°$) which satisfy the equation

$$3 \csc^2 \theta = 10 \cot \theta$$

(b) If $t = \tan \theta/2$ prove that

i $\sin \theta = \dfrac{2t}{1+t^2}$ ii $\cos \theta = \dfrac{1-t^2}{1+t^2}$

(c) Using i above show that $\tan 15° = 2 - \sqrt{3}$.
Using ii above show that $\tan 22\frac{1}{2}° = \sqrt{2} - 1$.

U.E.I.

7   On the same axis draw the curves $y = 3 \sin x$ and $y = \sin 3x$ for values of x from $0°$ to $180°$.

Scale 2 cm to $30°$ for x, 2 cm to one unit for y.

From these two draw the graph of $y = 3 \sin x + \sin 3x$ between these limits.

Show that there are two points at which the function is a maximum and find the values of x at which these occur.

D.T.C.

8   Express $0.12 \sin \theta - 0.35 \cos \theta$ in the $R \sin (\theta - \alpha)$ form.

Find the greatest and least values of the function and determine the corresponding values of θ.

Find in radians the first value of θ above 13π for which

$$0.12 \sin \theta - 0.35 \cos \theta = 0$$

[Give the answer correct to two decimal places.]   *D.T.C.*

9 (a) Prove that

$$\frac{\cos 2\theta + \cos 6\theta}{\cos 4\theta} = \frac{\sin 3\theta - \sin \theta}{\sin \theta}$$

(b) The angles being in radians, find the values of t between 0 and 2π for which

$$\cos \frac{t}{2} \cos \frac{3t}{2} + 0.44 = 0$$

E.M.E.U.

10 (a) Prove that

$$\frac{\sin (x + \frac{1}{4}\pi) - \cos (x + \frac{1}{4}\pi)}{\sin (x + \frac{1}{4}\pi) + \cos (x + \frac{1}{4}\pi)} = \tan x$$

(b) Express $24 \sin x + 7 \cos x$ in the form $R \sin (x + \theta)$ and hence find the angles between $0°$ and $360°$ which satisfy the equation

$$24 \sin x + 7 \cos x = -15$$

D.T.C.

11 (a) Assuming the formula

$$\cos (\theta + \phi) = \cos \theta \cos \phi - \sin \theta \sin \phi$$

find expressions for $\cos 2\theta$ and $\cos 3\theta$ in terms of $\cos \theta$.

(b) By making the substitution $x = 2 \cos \theta$, find the three roots of the equation

$$x^3 - 3x - 1 = 0$$

giving each root to three decimal places.

D.T.C.

12 Express $\cos \theta + 2 \cos (\theta + 60°)$ in the form $R \cos (\theta + \alpha)$, evaluating R and α. Deduce the maximum value of the expression and the smallest positive value of θ for which the expression takes this value. Find the values of θ between $0°$ and $360°$, for which $\cos \theta + 2 \cos (\theta + 60°) = 1$.

E.M.E.U.

13 Find all the solutions between $0°$ and $180°$ of the equations:

(a) $\sin (x + 15°) + \sin (3x + 135°) = 0$
(b) $\cos 4x - \cos 2x = \sin 4x - \sin 2x$

E.M.E.U.

14 (a) Write down the expression for $\sin \alpha + \sin \beta$ as a product. Show that

$$\sin 2\theta + 2 \sin 4\theta + \sin 6\theta = 16 \sin \theta \cos^3 \theta \cos 2\theta$$

(b) Express $\cos \theta + \sqrt{3} \sin \theta$ in the form $r \cos (\theta - \alpha)$, and find the value of θ between $0°$ and $90°$ for which

$$\cos \theta + \sqrt{3} \sin \theta = 1.2.$$

D.T.C.

147   Miscellaneous exercises 7

15 Express $1.2 \sin \theta - 0.5 \cos \theta$ in the form $R \sin (\theta - \alpha)$ and find the values of R (as a positive number) and α (in degrees), from this result.

(a) Solve the equation $1.2 \sin \theta - 0.5 \cos \theta = 0.65$ for values of θ between $0°$ and $360°$.

(b) Sketch the graph of $y = 1.2 \sin \theta - 0.5 \cos \theta$ (do not plot on squared paper) for a complete period, showing clearly those values of y and θ where the curve crosses the axes.

N.C.T.E.C.

16 Express the function $y = 3 \cos x - 4 \sin x$ in the form $R \cos (x + \alpha)$ and hence plot a graph of the function from $x = -\dfrac{\pi}{2}$ to $x = 2\pi$.

State the values of x (in radians) which make the function
(a) a maximum
(b) a minimum
(c) zero
(d) equal to 2.5

U.E.I.

17 (a) If $\tan 2A = \dfrac{3.0}{1.5 - b}$ and $\tan A = 2b$, find the value of b and the smallest possible value of A.

(b) Given $3 \sin (2x + y) + 4 \cos (2x + y) = 2.5$ and $x + y = \dfrac{\pi}{3}$, find the least positive values of x and y.

U.L.C.I.

18 (a) If θ is an acute angle and $\tan \theta = \frac{3}{4}$, find, without using tables, the value of $3 \sin 2\theta + 4 \cos 2\theta$.

(b) Solve the equation $\tan x \tan 2x = 1$ giving all the solutions between $0°$ and $360°$.

(c) Express $7 \sin \theta + 24 \cos \theta$ in the form $A \sin (\theta + \alpha)$ giving the values of A and α. Hence, find the maximum value of $7 \sin \theta + 24 \cos \theta$ and the value of θ at which the maximum occurs.

D.T.C.

19 (a) Express $\cos 2\theta - \cos 5\theta$ and $\sin 5\theta - \sin 2\theta$ each as a product and hence show that $\dfrac{\cos 2\theta - \cos 5\theta}{\sin 5\theta - \sin 2\theta} = \tan \dfrac{7\theta}{2}$.

(b) Express $16 \sin \theta - 21 \cos \theta$ in the form $R \sin (\theta - \phi)$ and hence find the value of θ to give a maximum value to $16 \sin \theta - 21 \cos \theta$.

U.L.C.I.

20 (a) If $y = 3 \sin (x + 30°) + 2 \cos (x + 60°)$, express y in the form $a \sin x + b \cos x$ and then convert this to the form $r \sin (x + a)$.

(b) Prove that for all values of A:

$$\frac{\tan (45° + \frac{1}{2}A)}{\tan (45° - \frac{1}{2}A)} = \frac{1 + \sin A}{1 - \sin A}.$$

D.T.C.

8 Methods of integration

8a Integration

In volume 1, chapter 12 integration was introduced as the reverse process to that of differentiation. Integration is also a summation, as will be shown later. The symbol for integration is the integral sign, used as follows:

$$\text{If } \frac{dy}{dx} = f(x) \text{ then } y = \int f(x)\, dx$$

$$\text{e.g. if } \frac{dy}{dx} = x^3 + 2x \text{ then } y = \int (x^3 + 2x)\, dx = \frac{x^4}{4} + x^2 + c$$

where c is an arbitrary constant. This result is true as it can be checked by differentiation, as can all integrations.

$$\text{Thus } \frac{d}{dx}\left(\frac{x^4}{4} + x^2 + c\right) = x^3 + 2x$$

From a knowledge of differentiation we obtain the following results which are fundamental and should be committed to memory.

$$\frac{d}{dx} x^n = nx^{n-1} \qquad\qquad \int x^n\, dx = \frac{1}{n+1} x^{n+1} + c$$
$$\text{except for } n = -1.$$

$$\frac{d}{dx} \log_e x = \frac{1}{x} \qquad\qquad \int \frac{dx}{x} = \log_e x + c$$

$$\frac{d}{dx} e^{ax} = ae^{ax} \qquad\qquad \int e^{ax}\, dx = \frac{1}{a} e^{ax} + c$$

$$\frac{d}{dx} \sin ax = a \cos ax \qquad\qquad \int \cos ax\, dx = \frac{1}{a} \sin ax + c$$

$$\frac{d}{dx} \cos ax = -a \sin ax \qquad\qquad \int \sin ax\, dx = -\frac{1}{a} \cos ax + c$$

$$\frac{d}{dx} \tan x = \sec^2 x \qquad\qquad \int \sec^2 x\, dx = \tan x + c$$

$$\frac{d}{dx} \cot x = -\operatorname{cosec}^2 x \qquad\qquad \int \operatorname{cosec}^2 x\, dx = -\cot x + c$$

In differentiating a sum of several terms each term is differentiated separately. Hence, reversing the process, in integrating a sum, each term is integrated separately.

Thus if u, v, w, etc. are functions of x then

$$\int (u+v-w+\ldots)\,dx = \int u\,dx + \int v\,dx - \int w\,dx + \ldots$$

Also a constant factor in the integrand (the expression to be integrated) can be written before the integral sign.

Thus $\int cy\,dx = c\int y\,dx$ where c is a constant.

The proof of this rule and of the one above is to differentiate and show that the derivatives of each side are equal.

Example Integrate the following

i $\dfrac{(1+\sqrt{x})^2}{x}$ ii $\cos 3x + \sin 2x$ iii $\frac{1}{2}(e^{2x}+e^{-2x})$

i $\displaystyle\int \frac{(1+\sqrt{x})^2}{x}\,dx = \int \frac{1+2\sqrt{x}+x}{x}\,dx$

$$= \int \left(\frac{1}{x}+2x^{-\frac{1}{2}}+1\right)dx = \log x + \frac{2x^{\frac{1}{2}}}{\frac{1}{2}}+x+c$$

$$= \log x + 4\sqrt{x}+x+c$$

ii $\displaystyle\int (\cos 3x + \sin 2x)\,dx = \frac{\sin 3x}{3}-\frac{\cos 2x}{2}+c$

iii $\int \frac{1}{2}(e^{2x}+e^{-2x})\,dx = \frac{1}{2}\left(\dfrac{e^{2x}}{2}+\dfrac{e^{-2x}}{-2}\right)+c$

$$= \tfrac{1}{4}(e^{2x}-e^{-2x})+c$$

Exercise 8a

Integrate the following with respect to the variable.

1 $x^2+3-\dfrac{1}{x^2}$ 2 $\sqrt{x}+\dfrac{1}{\sqrt{x}}$ 3 $\dfrac{1}{x^3}+x^3$

4 $(x+1)^2$ 5 $\dfrac{4x^3-3\sqrt{x}}{x}$ 6 $(1+\sqrt{x})^2$

7 $5x^{\frac{3}{2}}-x^{\frac{1}{2}}$ 8 $e^{2x}-\dfrac{1}{x}$ 9 $\dfrac{4}{x}-\dfrac{1}{x^2}$

10    $\dfrac{e^{3x}-e^{-3x}}{2}$ 11 $\cos 4x$ 12 $\sec^2 x - \sin x$

13    $e^{0\cdot2x}+x$ 14 $\cos 3x - \sin 2x$ 15 $\sin \dfrac{x}{2}$

16    $e^x + \dfrac{2}{x} - \dfrac{3}{e^x}$ 17 $4\sec^2 x - \operatorname{cosec}^2 x$ 18 $\dfrac{1}{2x}$

19    $\sin ax + \cos bx$ 20 $\left(x+\dfrac{1}{x}\right)^2$

21    $\dfrac{t+1}{\sqrt{t}}$ 22 $\cos 2\theta + \sin 3\theta$

23    $\sec^2 \theta + \cos 3\theta$ 24 $\dfrac{1}{v^{1\cdot4}} - \dfrac{1}{v}$

25    $e^t - \dfrac{1}{e^{2t}} + \dfrac{1}{t}$ 26 $\sin 100\pi t - \cos 50\pi t$

27    $\sqrt{r}(1+r)$ 28 $(t+1)(3-t)$

29    $(2+\sqrt{t})^2$ 30 $\operatorname{cosec}^2 x - \sin x$

8b   The definite integral

The expression $\int_a^b f(x)\,dx$ is used to represent the value of $\int f(x)\,dx$ when $x = b$ *minus* the value of $\int f(x)\,dx$ when $x = a$. Such an expression is called a definite integral because the arbitrary constant automatically disappears in the subtraction.

Example   Evaluate the definite integrals

i $\displaystyle\int_0^4 (x^2 - \sqrt{x})\,dx$ ii $\displaystyle\int_{\frac{1}{2}}^1 \cos x\,dx$ iii $\displaystyle\int_0^2 \dfrac{dx}{e^x}$

i $\displaystyle\int_0^4 (x^2 - x^{\frac{1}{2}})\,dx = \left[\dfrac{x^3}{3} - \dfrac{x^{\frac{3}{2}}}{\frac{3}{2}}\right]_0^4 = \dfrac{64}{3} - \dfrac{16}{3} = 16$

151   **The definite integral**

ii $\int_{\frac{1}{2}}^{1} \cos x \, dx = \left[\sin x \right]_{\frac{1}{2}}^{1} = \sin 1 - \sin \frac{1}{2}$ (where the angles are in radians)

$$= \sin 57 \cdot 3° - \sin 28 \cdot 65°$$
$$= 0 \cdot 8415 - 0 \cdot 4794$$
$$= 0 \cdot 362$$

iii $\int_{0}^{2} \dfrac{dx}{e^x} = \int_{0}^{2} e^{-x} \, dx = \left[-e^{-x} \right]_{0}^{2} = (-e^{-2} + e^{0})$

$$= 1 - 0 \cdot 1353 = 0 \cdot 8647 = 0 \cdot 865$$

Note that $\displaystyle\int_{a}^{b} f(x) \, dx = -\int_{b}^{a} f(x) \, dx$

for if $\int f(x) \, dx = F(x)$ then obviously

$$\int_{a}^{b} f(x) \, dx = F(b) - F(a) = -[F(a) - F(b)] = -\int_{b}^{a} f(x) \, dx$$

Thus the limits of the definite integral may be interchanged if the sign of the integral is changed.

8c Application to area under a curve

It was shown in volume 1 that $\displaystyle\int_{a}^{b} y \, dx$ equals the area bounded by the graph of y, the axis of x and the ordinates at $x = a$ and $x = b$. This important result is proved as follows:

The graph of $y = f(x)$ is represented in fig. 39.

Let P, coordinates x and y, be any point on the arc BC and let A represent the area BPNL.

If the ordinate PN moves a small distance δx to the position QR then area A will increase by the small amount PQRN which we denote by δA.

Ordinate QR $= y + \delta y$

δA lies between the values of the areas of two rectangular strips, each of width δx, one of height y and the other of height $y + \delta y$.

Figure 39

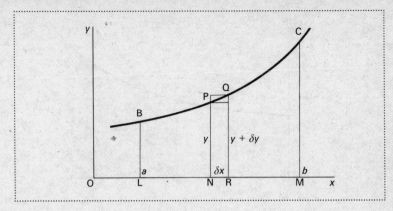

δA lies between $y\delta x$ and $(y+\delta y)\delta x$

$\dfrac{\delta A}{\delta x}$ lies between y and $y+\delta y$

As $\delta x \to 0$, $y+\delta y \to y$ and $\dfrac{\delta A}{\delta x} \to \dfrac{dA}{dx}$

Hence $\quad \lim\limits_{\delta x \to 0} \dfrac{\delta A}{\delta x} = \dfrac{dA}{dx} = y$

$$A = \int y \, dx$$

Let the value of this integral be $F(x)+C$.
Since the area A is measured starting at $x = a$,

then $\quad A = 0$ when $\quad x = a$
$\qquad 0 = F(a)+C \quad$ and $\quad C = -F(a)$
When $\quad x = b \qquad$ area BCML $= F(b)+C$
$\qquad\qquad\qquad\qquad\qquad\quad = F(b)-F(a)$
$\qquad\qquad\qquad\qquad\qquad\quad = \int\limits_{a}^{b} y \, dx$

Thus $\int\limits_{a}^{b} y \, dx$ equals the area bounded by the graph of y, the axis of x and the ordinates at $x = a$ and $x = b$.

Example Calculate the area bounded by the curve $y = \dfrac{e^x + e^{-x}}{2}$, the x-axis and the ordinates at $x = 1$ and $x = 2$.

Figure
40

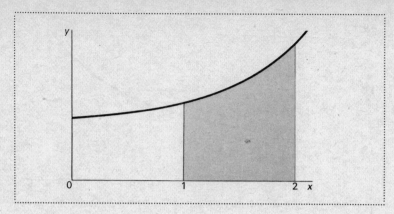

Required area
(shown shaded in fig. 40) $= \dfrac{1}{2} \int\limits_{1}^{2} (e^x + e^{-x})\, dx$

$$= \frac{1}{2} \left[e^x - e^{-x} \right]_{1}^{2}$$

$$= \frac{1}{2} \left(e^2 - \frac{1}{e^2} \right) - \frac{1}{2} \left(e - \frac{1}{e} \right)$$

$$= \tfrac{1}{2}(7\cdot3891 - 0\cdot1353) - \tfrac{1}{2}(2\cdot7183 - 0\cdot3679)$$

$$= 2\cdot45$$

Exercise 8c

Evaluate the following definite integrals.

1 $\int\limits_{0}^{2} x^2\, dx$

2 $\int\limits_{0}^{3} (4 - x)^2\, dx$

3 $\int\limits_{1}^{2} \dfrac{dx}{x^2}$

4 $\int\limits_{0}^{4} 3\sqrt{x}\, dx$

5 $\int\limits_{1}^{2} (t^2 + 3t)\, dt$

6 $\int\limits_{1}^{9} \dfrac{dp}{\sqrt{p}}$

7 $\int\limits_{1}^{2} \left(r^3 - \dfrac{1}{r} \right) dr$

8 $\int\limits_{0}^{\frac{1}{2}\pi} 4 \cos x\, dx$

9    $\int_0^{\frac{1}{4}\pi} (\cos x - \sin x)\, dx$

10    $\int_0^{\frac{1}{4}\pi} (2\cos x - \sin x)\, dx$

11    $\int_{\frac{1}{3}}^{\frac{1}{2}} \cos 2x\, dx$

12    $\int_0^{0\cdot 8} e^{2x}\, dx$

13    $\int_2^3 \dfrac{dx}{x}$

14    $\int_0^1 (e^{0\cdot 3x} + x^2)\, dx$

15    $\int_{\frac{1}{6}\pi}^{\frac{1}{3}\pi} \sec^2\theta\, d\theta$

16    $\int_0^{\frac{1}{4}\pi} (\cos 2\theta - \sin 3\theta)\, d\theta$

17    $\int_{\frac{1}{4}\pi}^{\frac{1}{3}\pi} \operatorname{cosec}^2\theta\, d\theta$

18    $\int_1^4 \dfrac{x+1}{2\sqrt{x}}\, dx$

19    $\int_{\frac{1}{2}}^1 \dfrac{e^t + e^{-t}}{2}\, dt$

20    $\int_0^{\pi/\omega} \sin \omega t\, dt$

21    $\int_1^4 \sqrt{t}(1+t)^2\, dt$

22    $\int_0^1 (\sqrt{x}+1)^2\, dx$

23    $\int_1^3 \left(\sqrt{t} + \dfrac{1}{\sqrt{t}}\right)^2\, dt$

24    $\int_{0\cdot 2}^{0\cdot 5} \cos 3x\, dx$

25    $\int_0^1 \sin 4x\, dx$

26    $\int_0^{\frac{1}{10}} \sin 5t\, dt$

27    $\int_1^2 (\cos\theta + \sin\theta)\, d\theta$

28    $\int_4^8 \dfrac{3\, dx}{x}$

29    $\int_3^4 (x+1)(2-x)\, dx$

30    $\int_1^2 \left(e^{2x} + \dfrac{1}{2x}\right)\, dx$

Find the area bounded by the given curve, the x-axis and the given ordinates in the following examples.

31 $y = x^2 - 2x$ $x = 0, x = 2$

32 $y = 3x + x^2$ $x = 1, x = 3$

33 $y = \sin x$ $x = 0, x = \dfrac{\pi}{4}$

34 $y = 3\cos 2x$ $x = \dfrac{\pi}{6}, x = \dfrac{\pi}{3}$

35 $y^2 = 4x$ $x = 1, x = 4$

36 $y = \dfrac{e^x - e^{-x}}{2}$ $x = 0, x = 2$

37 $xy = 4$ $x = 3, x = 6$

38 $y = \sin x + \cos x$ $x = \dfrac{\pi}{6}, x = 1$

39 $y = \dfrac{1}{e^{2x}}$ $x = 0, x = 2$

40 $xy^2 = 4$ $x = 1, x = 9$

To evaluate an integral it must be in one of the standard forms of the table in section 8a earlier in the chapter. The rest of this chapter is taken up with methods of integration by which some integrals may be rearranged so that they are in standard form.

8d Integration by substitution

If $\int f(x)\,dx$ is not recognizable as a standard integral we may try changing the variable x by a suitable substitution in terms of another variable, say u.

The integral $\int x\sqrt{(x^2+1)}\,dx$ may be solved by substituting $x^2 + 1 = u$.

We have to find a substitution for the term dx, which is known as a differential.

If $y = x^3$ instead of writing $\dfrac{dy}{dx} = 3x^2$ it is written $dy = 3x^2\,dx$. The terms

dy and dx when used thus are called *differentials*. Thus $dy = \dfrac{dy}{dx} \times dx$.

This notation is used in evaluating integrals by substitution.

For example, if $u = x^3 + 2x$

then $\qquad\qquad du = \dfrac{du}{dx} \times dx = (3x^2 + 2)\, dx$

If $u = \sin x$ then $du = \cos x\, dx$.

Example 1 Find the value of $\int x\sqrt{(x^2 + 1)}\, dx$.

Let $\quad x^2 + 1 = u \quad$ then $\quad du = \dfrac{du}{dx} \times dx$

$$= 2x\, dx$$

$$\int x\sqrt{(x^2 + 1)}\, dx = \tfrac{1}{2} \int (x^2 + 1)^{\frac{1}{2}} \times 2x\, dx$$

$$= \tfrac{1}{2} \int u^{\frac{1}{2}}\, du$$

$$= \frac{u^{\frac{3}{2}}}{2 \times \frac{3}{2}} + c$$

$$= \tfrac{1}{3} u^{\frac{3}{2}} + c\,.$$

Since $\qquad u = x^2 + 1 \quad$ then

$$\int x\sqrt{(x^2 + 1)}\, dx = \tfrac{1}{3}(x^2 + 1)^{\frac{3}{2}} + c$$

Example 2 Evaluate $\int (6x - 7)^8\, dx$.

Let $\quad u = 6x - 7,\ du = 6dx$

$$\int (6x - 7)^8\, dx = \frac{1}{6} \int u^8\, du = \frac{1}{6 \times 9} u^9 + c$$

$$= \tfrac{1}{54}(6x - 7)^9 + c$$

Example 3 Evaluate $\int \cos (3x - 1)\, dx$.

Let $\quad 3x - 1 = u \quad$ then $\quad du = 3dx$

$$\int \cos (3x - 1)\, dx = \int \cos u\, \frac{du}{3} = \frac{\sin u}{3} + c$$

$$= \tfrac{1}{3} \sin (3x - 1) + c$$

Example 4 Find the value of $\int \dfrac{dx}{3x+1}$.

Let $u = 3x+1$ then $du = 3dx$

Substituting in the integral gives

$$\int \frac{dx}{3x+1} = \int \frac{1}{u} \frac{du}{3} = \frac{1}{3} \int \frac{du}{u}$$

$$= \tfrac{1}{3} \log u + c$$

$$= \tfrac{1}{3} \log (3x+1) + c$$

Example 5 Evaluate $\int \sin^7 x \cos x \, dx$.

Let $u = \sin x, \, du = \cos x \, dx$

$$\int \sin x^7 \cos x \, dx = \int u^7 \, du = \frac{u^8}{8} + c$$

$$= \tfrac{1}{8} \sin^8 x + c$$

Exercise 8d

By using suitable substitutions find the value of the following integrals.

1   $\int (x+4)^2 \, dx$
2   $\int (x+1)^{10} \, dx$

3   $\int (2-x)^3 \, dx$
4   $\int \sqrt{(2x+3)} \, dx$

5   $\int \dfrac{dx}{(2x+1)^3}$
6   $\int \dfrac{dx}{2x+1}$

7   $\int e^{2x} \, dx$
8   $\int e^{-3x} \, dx$

9   $\int \sin 3x \, dx$
10   $\int \cos 4x \, dx$

11   $\int (x^2+1)^5 \, 2x \, dx$
12   $\int (x^2+1)^3 \, x \, dx$

13   $\int \sqrt{(x^2+x)}(2x+1) \, dx$
14   $\int (ax^2+bx)^4(2ax+b) \, dx$

15 $\int \cos(3x+2)\,dx$

16 $\int (2x+1)\cdot\cos(x^2+x)\,dx$

17 $\int \sin(4-x)\,dx$

18 $\int \dfrac{x^5\,dx}{(x^6+1)^2}$

19 $\int e^{5x^2-x}(10x-1)\,dx$

20 $\int \dfrac{x^3}{\sqrt{(a^4-x^4)}}\,dx$

21 $\int \cos^3 x \sin x\,dx$

22 $\int \sin^2 x \cos x\,dx$

23 $\int \tan x \sec^2 x\,dx$

24 $\int \dfrac{\log x}{x}\,dx$

25 $\int \cos(2-3x)\,dx$

26 $\int e^{-x^2} x\,dx$

27 $\int \dfrac{dx}{e^{3x}}$

28 $\int \sin(3-x)\,dx$

29 $\int \sin^5 x \cos x\,dx$

30 $\int \dfrac{e^x}{1+e^x}\,dx$

31 $\displaystyle\int_0^1 \cos\left(\dfrac{x}{2}+1\right)dx$

32 $\displaystyle\int_0^{\frac{1}{2}} (2x+1)^4\,dx$

33 $\displaystyle\int_0^{\frac{1}{2}\pi} \sin(3x+1)\,dx$

34 $\displaystyle\int_0^1 (2x-1)^7\,dx$

35 $\displaystyle\int_{-1}^2 (3x+4)\,dx$

36 $\displaystyle\int_3^4 \dfrac{dx}{x-2}$

37 $\displaystyle\int_0^{0.3} \sin(2x+1)\,dx$

38 $\displaystyle\int_2^3 \dfrac{dx}{(x-1)^2}$

39 $\displaystyle\int_0^{\frac{1}{2}\pi} \sin^3 x \cos x\,dx$

40 $\displaystyle\int_3^5 3x\sqrt{(x^2-9)}\,dx$

159   **Integration by substitution**

8e Integration using partial fractions

Most fractions, if the denominator is a product of linear factors, are integrable by splitting them into partial fractions.

Example 1 Evaluate $\int \dfrac{3x+4}{(x-2)(x+3)}\,dx$.

Let $\dfrac{3x+4}{(x-2)(x+3)} \equiv \dfrac{A}{x-2} + \dfrac{B}{x+3}$ where A and B are constants.

$$3x+4 \equiv A(x+3) + B(x-2) \quad \text{true for all values of } x.$$

When $x = 2$ $\quad 6+4 = A \times 5$
$$A = 2$$
When $x = -3$ $\quad -9+4 = B(-5)$
$$B = 1$$

$$\int \dfrac{3x+4}{(x-2)(x+3)}\,dx = \int \left(\dfrac{2}{x-2} + \dfrac{1}{x+3} \right) dx$$

$$= 2\int \dfrac{dx}{x-2} + \int \dfrac{dx}{x+3}$$

$$= 2 \log (x-2) + \log (x+3) + C$$

$$= \log (x-2)^2 (x+3) + C$$

Example 2 Evaluate $\displaystyle\int_{4}^{6} \dfrac{x^2+2}{x-2}\,dx$.

$\dfrac{x^2+2}{x-2}$ is not a proper fraction and therefore the numerator must be divided by the denominator until the degree of the remainder is less than that of the denominator.

$$\dfrac{x^2+2}{x-2} = x+2+\dfrac{6}{x-2}$$

$$\int_{4}^{6} \dfrac{x^2+2}{x-2}\,dx = \int_{4}^{6} \left(x+2+\dfrac{6}{x-2} \right) dx$$

$$= \left[\dfrac{x^2}{2} + 2x + 6 \log (x-2) \right]_{4}^{6}$$

$$= 18+12+6 \log 4 - (8+8+6 \log 2)$$

$$= 14+6 \log 2 = 14+6 \times 0{\cdot}6931$$

$$= 18{\cdot}2$$

Exercise 8e

Integrate

1 $\dfrac{x}{x+3}$

2 $\dfrac{x^2}{x+3}$

3 $\dfrac{x}{2x-3}$

4 $\dfrac{2x+3}{x-4}$

5 $\dfrac{2x-5}{x^2-5x+6}$

6 $\dfrac{2x+3}{x^2+x-30}$

7 $\dfrac{x^2}{x^2-4}$

8 $\dfrac{x+1}{3x^2-x-2}$

9 $\dfrac{x+1}{(x-1)^2}$

10 $\dfrac{5x+2}{x^2-4x+4}$

11 $\dfrac{x+3}{x(2x+3)}$

12 $\dfrac{x^2-3x+3}{(x-1)(x-2)(x-3)}$

13 $\dfrac{1}{x^2-1}$

14 $\dfrac{\cos x}{\sin x \,(1+\sin x)}$ (put $u = \sin x$)

15 $\dfrac{1}{4x^2-9}$

16 $\dfrac{1}{4-x^2}$

17 $\dfrac{e^x}{1-e^{2x}}$ (put $u = e^x$)

18 $\dfrac{\sin x}{1-4\cos^2 x}$ (put $u = \cos x$)

19 $\dfrac{1}{x^2-2x-1}$

20 $\dfrac{1}{x^2-4x-8}$

21 $\dfrac{1}{x^2-4x-12}$

22 $\dfrac{1}{4-2x-x^2}$

8f Integration by parts

From the formula for differentiating a product,

$$\frac{d}{dx}(uv) = u\frac{dv}{dx} + v\frac{du}{dx}$$

$$u\frac{dv}{dx} = \frac{d}{dx}(uv) - v\frac{du}{dx}$$

161 Integration by parts

Integrating both sides with respect to x,

$$\int u \frac{dv}{dx} dx = uv - \int v \frac{du}{dx} dx$$

or more simply

$$\int u \, dv = uv - \int v \, du$$

This result is used when the integrand is the product of two functions and it is easier to find $\int v \, du$ than $\int u \, dv$.

Example 1   Evaluate $\int xe^x \, dx$.

Here we try $u = x$ and $dv = e^x \, dx$

$$v = \int e^x \, dx = e^x$$

Thus $\int xe^x \, dx = xe^x - \int e^x \, dx$

$$= xe^x - e^x - c$$

Example 2   Find $\int x \cos x \, dx$.

Let $x = u, \cos x \, dx = dv$
$$v = \sin x$$

$\int x \cos x \, dx = x \sin x - \int \sin x \, dx$

$$= x \sin x + \cos x + c$$

Example 3   Evaluate $\int_1^3 x^3 \log x \, dx$.

Let $\log x = u$ and $x^3 \, dx = dv$

Then $v = \dfrac{x^4}{4}$ and $du = \dfrac{d}{dx} (\log x) \times dx = \dfrac{1}{x} dx$

$$\int_1^3 x^3 \log x \, dx = \left[\frac{x^4}{4} \times \log x - \int \frac{x^4}{4} \times \frac{1}{x} dx \right]_1^3$$

$$\int_1^3 x^3 \log x \, dx = \left[\frac{x^4 \log x}{4} - \frac{1}{4} \int x^3 \, dx \right]_1^3$$

$$= \left[\frac{x^4 \log x}{4} - \frac{x^4}{16} \right]_1^3$$

$$= \frac{81 \log_e 3}{4} - \frac{81}{16} - \left(0 - \frac{1}{16} \right)$$

$$= \frac{81 \log_e 3}{4} - 5 = 17 \cdot 2$$

Exercise 8f

Using integration by parts, find the values of the integrals in examples 1 to 10.

1   $\int \log x \, dx$ (take $u = \log x$ and $dv = dx$)   2 $\int x \log x \, dx$

3   $\int \frac{\log x}{x^3} \, dx$ 4 $\int x \cos 3x \, dx$ 5 $\int x \sin x \, dx$

6   $\int x(1+x)^{10} \, dx$ 7 $\int x^7 \log x \, dx$ 8 $\int \sqrt{x} \log x \, dx$

9   $\int x^2 e^x \, dx$ 10 $\int x^2 \cos x \, dx$

Evaluate the integrals

11   $\int_0^3 x e^x \, dx$ 12 $\int_1^2 t e^{-1} \, dt$ 13 $\int_1^3 x^2 \log x \, dx$

14   $\int_0^{\frac{\pi}{2}} t \cos t \, dt$ 15 $\int_1^2 x e^{2x} \, dx$

8g   Integration of trigonometric functions

Some trigonometric integrals have already been evaluated by the method of substitution as in the following example.

Example 1 Integrate i $\sin(ax+b)$, ii $\sin^m x \cos x$.

i Use the substitution $u = ax+b$, $du = a\,dx$.

$$\int \sin(ax+b)\,dx = \frac{1}{a}\int \sin u\,du = -\frac{\cos u}{a} = -\frac{\cos(ax+b)}{a}+c$$

ii Use the substitution $u = \sin x$, $du = \cos x\,dx$.

$$\int \sin^m x \cos x\,dx = \int u^m\,du = \frac{u^{m+1}}{m+1} = \frac{\sin^{m+1} x}{m+1}+c$$

Similarly $\quad \int \cos^n x \sin x\,dx = -\frac{\cos^{n+1} x}{n+1}+c$

Identities are very useful in changing trigonometric functions into forms suitable for integration.

Example 2 Integrate i $\sin^2 x$, ii $\cos^2 x$, iii $\sin 5x \cos 3x$, iv $\cos 6x \cos 2x$.

i and ii The identity $\cos 2x = 2\cos^2 x - 1 = 1 - 2\sin^2 x$

gives $\quad \sin^2 x = \frac{1}{2}(1-\cos 2x)$
and $\quad \cos^2 x = \frac{1}{2}(1+\cos 2x)$

$$\int \sin^2 x\,dx = \frac{1}{2}\int(1-\cos 2x)\,dx = \frac{1}{2}\left(x-\frac{\sin 2x}{2}\right)+c$$

$$\int \cos^2 x\,dx = \frac{1}{2}\int(1+\cos 2x)\,dx = \frac{1}{2}\left(x+\frac{\sin 2x}{2}\right)+c$$

iii Using $\quad \sin \times \cos = \frac{1}{2}[\sin(\text{sum}) + \sin(\text{difference})]$

then $\quad \sin 5x \cos 3x = \frac{1}{2}[\sin 8x + \sin 2x]$

$$\int \sin 5x \cos 3x\,dx = \frac{1}{2}\int(\sin 8x + \sin 2x)\,dx$$

$$= \frac{1}{2}\left(-\frac{\cos 8x}{8} - \frac{\cos 2x}{2}\right)+c$$

iv $\int \cos 6x \cos 2x\,dx = \frac{1}{2}\int(\cos 8x + \cos 4x)\,dx$

$$= \frac{1}{2}\left(\frac{\sin 8x}{8} + \frac{\sin 4x}{4}\right)+c$$

Some difficult integrals may be solved by suitable trigonometric substitutions.

Example 3 Evaluate $\displaystyle\int \frac{dx}{\sqrt{(a^2-x^2)}}$ by using the substitution $x = a \sin \theta$.

164 Methods of integration

Putting $x = a \sin \theta$, $dx = a \cos \theta \, d\theta$.

$$\sqrt{(a^2 - x^2)} = \sqrt{(a^2 - a^2 \sin^2 \theta)} = a\sqrt{(1 - \sin^2 \theta)} = a \cos \theta$$

$$\int \frac{dx}{\sqrt{(a^2 - x^2)}} = \int \frac{a \cos \theta \, d\theta}{a \cos \theta} = \int d\theta = \theta + c \quad \text{where } c \text{ is a constant.}$$

Since $x = a \sin \theta$, $\sin \theta = \dfrac{x}{a}$ and $\theta = \sin^{-1} \dfrac{x}{a}$.

Thus $\displaystyle \int \frac{dx}{\sqrt{(a^2 - x^2)}} = \sin^{-1} \frac{x}{a} + c$

Example 4 Evaluate $\displaystyle \int_{1}^{2} \frac{dx}{4 + x^2}$ using the substitution $x = 2 \tan \theta$.

Let $x = 2 \tan \theta$, $dx = 2 \sec^2 \theta \, d\theta$.

$$4 + x^2 = 4 + 4 \tan^2 \theta = 4(1 + \tan^2 \theta) = 4 \sec^2 \theta$$

$$\int \frac{dx}{4 + x^2} = \int \frac{2 \sec^2 \theta}{4 \sec^2 \theta} \, d\theta = \frac{1}{2} \int d\theta = \frac{\theta}{2} + c$$

But $\tan \theta = \dfrac{x}{2}$ and $\theta = \tan^{-1} \dfrac{x}{2}$

$$\int_{1}^{2} \frac{dx}{4 + x^2} = \frac{1}{2} \left[\tan^{-1} \frac{x}{2} \right]_{1}^{2} = \frac{1}{2} \left(\tan^{-1} 1 - \tan^{-1} \frac{1}{2} \right)$$

$$= \frac{1}{2} \left(\frac{\pi}{4} - 0.4637 \right) = 0.161$$

Exercise 8g

Integrate the following trigonometric functions.

1   $\sin 6x$ 2 $\sin (2x - 1)$

3   $\cos (3x + 2)$ 4 $\cos (8t - 1)$

5   $\sin^2 3x$ 6 $\cos^2 \dfrac{t}{2}$

7   $\sin 6x \cos 5x$ 8 $\cos 3t \cos t$

9   $\sin 3t \sin t$ 10 $\sin 4x \sin 2x$

11   $\cos 2x \cos 5x$ 12 $\sin 2\theta \cos 5\theta$

13 $\sin \dfrac{t}{2} \cos \dfrac{3t}{2}$ 14 $\cos^4 \theta \sin \theta$

15    $\sin^5 t \cos t$ 16 $(\sin^4 \theta + 2) \cos \theta$

17    $\sin^3 \theta$ (put $\sin^2 \theta = 1 - \cos^2 \theta$)

18    $\dfrac{\cos x}{\sin^3 x}$ 19 $\dfrac{\sin \theta}{\cos \theta}$

20    $\tan^3 \theta \sec^2 \theta$

Evaluate the following.

21    $\displaystyle\int_0^{\frac{\pi}{6}} \sin^2 x \, dx$ 22 $\displaystyle\int_0^1 \cos(2x+1) \, dx$

23    $\displaystyle\int_0^{\frac{\pi}{2}} \sin 3x \sin 2x \, dx$ 24 $\displaystyle\int_0^{\frac{\pi}{3}} \sin^3 t \cos t \, dt$

25    $\displaystyle\int_0^2 \dfrac{dx}{\sqrt{(4-x^2)}}$ 26 $\displaystyle\int_0^1 \dfrac{dx}{1+x^2}$

Miscellaneous exercises 8

1   Illustrate geometrically, by representing the integral by the area under the graph, the following identities.

(a) $\displaystyle\int_{-a}^{a} f(x) \, dx = 2 \int_0^a f(x) \, dx$

when $f(x)$ is an 'even' function of x, i.e. the graph is symmetrical about the Oy axis and $f(-a) = f(a)$.

(b) $\displaystyle\int_{-a}^{a} f(x) \, dx = 0$

when $f(x)$ is an 'odd' function of x, i.e. the graph is skew symmetrical and $f(-a) = -f(a)$.

2   (a) Evaluate i $\displaystyle\int_1^4 \left(x^3 - \dfrac{1}{x^2} + x\right) dx$ ii $\displaystyle\int_0^{\frac{1}{3}\pi} \cos(3x+1) \, dx$

(b) Calculate the area bounded by the graph of $y = x^3$, the axis of x and the ordinates at $x = 1$ and $x = 2$.

3   Evaluate (a) $\int_1^4 \left(\sqrt{x} + \dfrac{2}{\sqrt{x}} \right) dx$ (b) $\int_0^{\frac{1}{6}\pi} \sin 3x\, dx$ (c) $\int_0^1 e^{2x}\, dx$

Find the area bounded by the curve $y = x^2 + 1$, the x-axis and the ordinates at $x = 1$ and $x = 2$.

4   (a) Evaluate   i $\int_1^2 \dfrac{x-1}{x+1}\, dx$ ii $\int_{\frac{1}{6}\pi}^{\frac{1}{3}\pi} \left[2\theta + \cos \left(2\theta - \dfrac{\pi}{3} \right) \right] d\theta$

(b) Show, using partial fractions, that

$$\int_3^8 \dfrac{7x\, dx}{(3x+1)(x-2)} = 2 \log_e 6 + \tfrac{1}{3} \log_e 2 \cdot 5$$

U.E.I.

5   Evaluate the following integrals

(a) $\int \dfrac{x}{1+x^2}\, dx$ (b) $\int (x^{\frac{1}{2}} + x^{-\frac{5}{4}})\, dx$ (c) $\int_1^2 \dfrac{3\, dx}{\sqrt{(4-x^2)}}$, by putting $x = 2 \sin \theta$.

D.T.C.

6   (a) Differentiate the following with respect to x and express your answers in a simple form.

   i $x\sqrt{(x^2-1)}$ ii $\dfrac{e^{2x} - e^{-2x}}{e^{2x} + e^{-2x}}$

(b) Evaluate the following correct to two decimal places.

   i $\int_0^2 \dfrac{1}{(4-x)}\, dx$ ii $\int_{-1}^1 (x^2 + 1)^2\, dx$

N.C.T.E.C.

7   Evaluate

(a) $\int_0^{\frac{1}{4}\pi} 2 \sin (\pi - 2x)\, dx$ (b) $\int_0^2 2x(x^2 - 4)\, dx$

(c) $\int_{0\cdot2}^{0\cdot4} \dfrac{1\, dx}{1 - 2x}$ (d) $\int_0^1 a e^{ax}\, dx$

U.L.C.I.

8 (a) Evaluate the integrals

$$\text{i} \int_0^{\frac{1}{2}\pi} \sin^2 x \cos x \, dx \qquad \text{ii} \int_0^2 \frac{x+1}{x^2+2x+2} \, dx \qquad \text{iii} \int_1^2 x \log_e x \, dx$$

(b) Evaluate the integral $\displaystyle\int_0^4 \frac{x}{\sqrt{(4-x)}} \, dx$ by means of the substitution $z^2 = 4-x$.

D.T.C.

9 (a) Obtain $\displaystyle\int \frac{dx}{3-x}$.

(b) Evaluate correct to three significant figures

$$\text{i} \int_{\frac{1}{6}\pi}^{\frac{1}{3}\pi} \sin 2x \, dx \qquad \text{ii} \int_1^4 \left(\sqrt{x}+\frac{1}{\sqrt{x}}\right)^2 dx \qquad \text{iii} \int_0^2 e^{0.5t} \, dt$$

10 Evaluate

(a) $\displaystyle\int_0^{\frac{1}{2}\pi} \sin x \cos x \, dx$

(b) $\displaystyle\int_0^1 \frac{dx}{x+\sqrt{x}}$, by the substitution $\sqrt{x} = u$

(c) $\displaystyle\int_0^{\frac{1}{6}\pi} x \cos\left(2x-\frac{\pi}{3}\right) dx$, by integration by parts.

D.T.C.

11 (a) Express $\displaystyle\frac{x}{x^2+5x+6}$ in partial fractions and show that

$$\int_{-1}^1 \frac{x \, dx}{x^2+5x+6} = \log_e \frac{8}{9}$$

(b) Show that $y = e^{-x} \cos x$ satisfies the equation

$$\frac{d^2y}{dx^2} + 2\frac{dy}{dx} + 2y = 0$$

D.T.C.

12 (a) Evaluate

i $\int\limits_{1}^{3}\left(x+\dfrac{1}{x^2}\right)x\,dx$ ii $\int\limits_{-1}^{1\cdot303}(e^{x+1}+1)\,dx$ iii $\int\limits_{0}^{\frac{1}{4}\pi}3\sec x\,(\cos x+\sec x)\,dx$

(b) Show that $\dfrac{10x}{6x^2-13x+6}=\dfrac{6}{2x-3}-\dfrac{4}{3x-2}$

and hence find $\int\dfrac{10x\,dx}{6x^2-13x+6}$

U.E.I.

13 Evaluate

(a) $\int\limits_{0}^{\frac{1}{3}\pi}\cos\left(3x+\dfrac{\pi}{6}\right)dx$ (b) $\int\limits_{1}^{2}(e^{2x}-e^{-2x})\,dx$

(c) $\int\limits_{0}^{\frac{1}{2}\pi}2\sin^5\theta\cos\theta\,d\theta$ (d) $\int\limits_{1}^{4}\left(\sqrt{x}+\dfrac{9}{\sqrt{x}}\right)dx$

14 (a) Evaluate i $\int\limits_{1}^{8}\left(\sqrt[3]{x}-\dfrac{1}{\sqrt[3]{x}}\right)dx$ ii $\int\limits_{0}^{\frac{1}{4}\pi}\sin x\cos x\,dx$

(b) Evaluate i $\int\dfrac{2x-5}{2x^2+3x-2}\,dx$ using partial fractions

ii $\int x\log_e 2x\,dx$, by integrating by parts.   *D.T.C.*

15 (a) State the formula for integration by parts and use it to show that

$\int\limits_{1}^{2}xe^x\,dx=e^2$

(b) Find $\int\dfrac{\sin x}{2+3\cos x}\,dx$.

D.T.C.

16 Obtain

(a) $\int\left(\sqrt[3]{x}-\dfrac{3}{x^2}\right)dx$ (b) $\int e^{2x-3}\,dx$

(c) $\int\dfrac{dx}{\sqrt{(2+3x)}}$ (d) $\int\cos 4x\sin 2x\,dx$

169   Miscellaneous exercises 8

17 (a) Find i $\int \dfrac{x^2+a^2}{\sqrt{x}}\,dx$ ii $\int \cos 5x \cos 3x\,dx$

(b) Express $\dfrac{2x^2}{1-x^2}$ in a form suitable for integration using the method of partial fractions.

Hence evaluate $\int_{\frac{1}{4}}^{\frac{1}{2}} \dfrac{2x^2}{1-x^2}\,dx$ using partial fractions.

18 (a) Evaluate the following definite integrals giving the answers correct to three significant figures.

i $\int_{\frac{1}{2}}^{2}\left(x+\dfrac{1}{x}\right)^2 dx$ ii $\int_0^1 e^{-2x}\,dx$ iii $\int_1^8 \dfrac{\sqrt[3]{x+1}}{x}\,dx$

(b) Express $\cos 2\theta$ in terms of $\cos \theta$. Hence evaluate $\int_0^{\frac{1}{4}\pi} \cos^2 \theta\,d\theta$ correct to three significant figures.

E.M.E.U.

19 Evaluate

(a) $\int_4^9 \dfrac{dx}{1+2x}$ (b) $\int_0^{\frac{1}{4}\pi} \sin\left(x+\dfrac{\pi}{2}\right)dx$

(c) $\int_0^6 e^{\frac{1}{3}x}\,dx$ (d) $\int \sin \dfrac{k\theta}{2} \cos \dfrac{k\theta}{2}\,d\theta$

U.L.C.I.

20 (a) Evaluate i $\int xe^{2x}\,dx$ using integration by parts

ii $\int_0^{\frac{1}{4}\pi} \dfrac{\sin x \cos x}{\sqrt{(1+2\sin^2 x)}}\,dx$ by putting $1+2\sin^2 x = u$.

(b) Taking whole number values for x, find the area under the curve $y = \dfrac{1}{x}$ between the ordinates at $x = 1$ and $x = 5$ using Simpson's rule. Find this area also by integration.

D.T.C.

9 Applications of integration

9a Integration as a summation

Consider the area under the graph of $y = f(x)$ from $x = a$ to $x = b$ as represented by ABCD in fig. 41.

Let this area be divided into a large number of narrow strips of width δx and let the ordinates be y_1, y_2, y_3, etc. as shown.

Figure 41

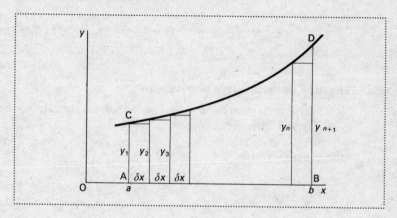

Then the area under the steps from C to D formed by the rectangles is

$$y_1\,\delta x + y_2\,\delta x + y_3\,\delta x + \ldots + y_n\,\delta x = \sum_1^n y\,\delta x$$

where $\sum$ = sum of.

As $\delta x \to 0$ and the number of strips approaches infinity then the area approaches the area under the curve CD.

This area has already been shown to equal the definite integral $\int_a^b y\,dx$.

Thus $\int_a^b y\,dx = \lim_{\delta x \to 0} \sum_{x=a}^{x=b} y\,\delta x$

This is the fundamental theorem of integration and it is applied to find volumes, centres of gravity, mean values, moments of inertia, etc. by taking a sample element and summing it over the region. Integration, previously defined as the reverse of differentiation, now becomes essentially the evaluation of the limit of a sum of elements.

Figure 42

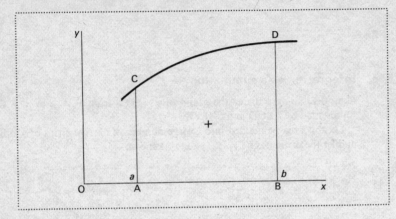

9b The area under a curve and its sign

Fig. 42 shows the portion of the graph $y = f(x)$ between ordinates $x = a$ and $x = b$. $f(x)$ is positive in this interval.

$$\text{Area} = \lim_{\delta x \to 0} \sum_{x=a}^{x=b} f(x)\, \delta x$$

$$= \int_a^b f(x)\, dx$$

Figure 4

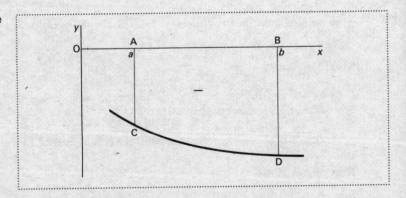

172   **Applications of integration**

As every term in $\sum f(x)\,\delta x$ is positive the area is positive and $\int\limits_a^b f(x)\,dx$ is positive.

However, if $f(x)$ is negative every term in $\sum f(x)\,\delta x$ is negative and $\int\limits_a^b f(x)\,dx$ is negative.

Figure 44

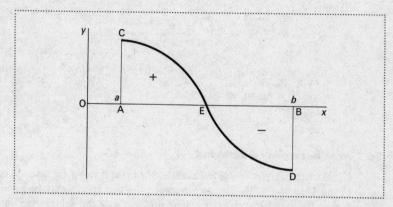

If the integrand is positive in part of the range and negative in the other part then $\int\limits_a^b f(x)\,dx = $ (area above x-axis) $-$ (area below x-axis).

Example 1 Find the area between the curve $y = x(x-1)(x-3)$ and the x-axis from $x = 0$ to $x = 3$.

$$y = x(x-1)(x-3)$$
$$= x^3 - 4x^2 + 3x$$

The graph has the typical cubic shape and is shown in fig. 45.

Figure 45

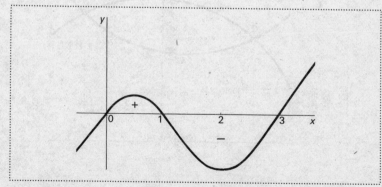

The area must be calculated in two parts since one area is positive and the other negative.

$$A_1 = \int_0^1 (x^3 - 4x^2 + 3x)\,dx = \left[\frac{x^4}{4} - \frac{4x^3}{3} + \frac{3x^2}{2} \right]_0^1 = \frac{5}{12}$$

$$A_2 = \int_1^3 (x^3 - 4x^2 + 3x)\,dx = \left[\frac{x^4}{4} - \frac{4x^3}{3} + \frac{3x^2}{2} \right]_1^3 = -2\tfrac{2}{3}$$

Area between curve and x-axis from $x = 0$ to $x = 3$

$$= \tfrac{5}{12} + 2\tfrac{2}{3} = 3\tfrac{1}{12} \text{ sq. units}$$

9c   Area between two curves

Consider the graphs of the functions $y = f_1(x)$ and $y = f_2(x)$ intersecting at C and D as in fig. 46. The area between the two curves may be found by finding the area between each graph and the x-axis and subtracting the two areas.

Thus   area $= \int_a^b f_2(x)\,dx - \int_a^b f_1(x)\,dx$

Alternatively, the area may be obtained by dividing it into a large number of rectangular strips of width δx and summing these elemental strips from $x = a$ to $x = b$.

Figure 46

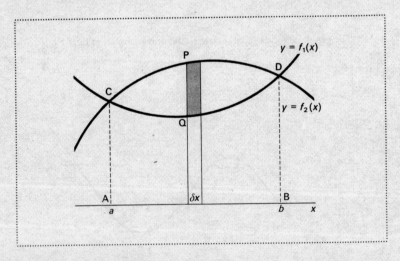

Area of element PQ (shown shaded in fig. 46) $= [f_2(x) - f_1(x)] \, \delta x$

$$\text{Total area} = \lim_{\delta x \to 0} \sum_{x=a}^{x=b} [f_2(x) - f_1(x)] \, \delta x$$

$$= \int_a^b [f_2(x) - f_1(x)] \, dx$$

Example Find the area between the curve $y = 6x - x^2$ and the line $y = x + 4$.

To find the points of intersection of the two graphs solve the equations by substituting $y = x + 4$ in $y = 6x - x^2$.

Thus $\qquad x + 4 = 6x - x^2$

i.e. $\quad x^2 - 5x + 4 = 0$

$\qquad (x - 1)(x - 4) = 0$

$x = 1, y = 5$, and $x = 4, y = 8$

The points are C (1, 5) and D (4, 8). See fig. 47.

Figure 47

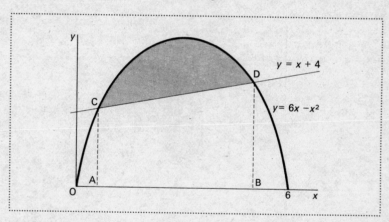

Required area $= \int_1^4 [(6x - x^2) - (x + 4)] \, dx$

$$= \int_1^4 (5x - x^2 - 4) \, dx = \left[\frac{5x^2}{2} - \frac{x^3}{3} - 4x \right]_1^4$$

$$= \left(40 - \frac{64}{3} - 16 \right) - \left(\frac{5}{2} - \frac{1}{3} - 4 \right)$$

$$= 4\tfrac{1}{2} \text{ sq. units}$$

9d Volume of a solid of revolution

When a plane area rotates to make a complete revolution about an axis in its plane it generates a solid known as a solid of revolution.

Any section of the solid perpendicular to the axis is a circle and the volume can be obtained as an integral.

1 Rotation about the x-axis

Let V be the volume generated when the area bounded by the graph of $y = f(x)$, the x-axis and the ordinates $x = a$ and $x = b$, makes one revolution about the x-axis.

Let the volume of this solid be divided into a number of thin discs by planes perpendicular to the x-axis.

Consider a sample element of radius y and thickness δx as shown in fig. 48.

Volume of element $= \pi y^2 \, \delta x$

Figure 48

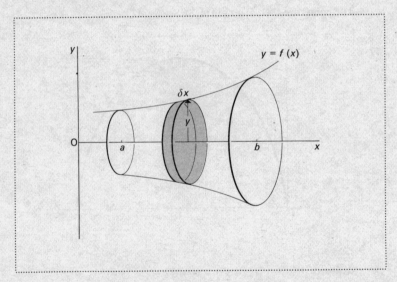

The total volume will be the sum of all such elements

$$= \lim_{\delta x \to 0} \sum_{x=a}^{x=b} \pi y^2 \, \delta x$$

$$V = \int_a^b \pi y^2 \, dx$$

2 Rotation about the y-axis

This is similar to the above with x and y interchanged. See fig. 49.

$$V = \int_{y_1}^{y_2} \pi x^2 \, dy$$

Figure 49

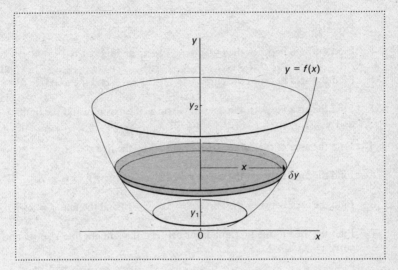

Example Find the volume of the solid of revolution obtained by rotating the curve $y = x^2$ about

  i the x-axis between the limits $x = 1$ and $x = 3$,
  ii the y-axis between the same limits.

  i Volume $= \int_1^3 \pi y^2 \, dx$

$$= \int_1^3 \pi x^4 \, dx = \pi \left[\frac{x^5}{5} \right]_1^3$$

$$= \pi \left(\frac{243}{5} - \frac{1}{5} \right) = 48\tfrac{2}{5}\pi$$

$$= 152 \text{ cubic units}$$

  ii When $x = 1$, $y = x^2 = 1$
         $x = 3$, $y = 3^2 = 9$

$$\text{Volume} = \int_1^9 \pi x^2 \, dy = \int_1^9 \pi y \, dy$$

$$= \pi \left[\frac{y^2}{2} \right]_1^9 = \pi \left(\frac{81}{2} - \frac{1}{2} \right) = 40\pi$$

$$= 125 \text{ cubic units}$$

Exercise 9d

1 Find the area between the curve $y = 2x - x^2$ and the x-axis.

2 Find the area between the curve $y = (3-x)(x-2)$ and the x-axis.

3 Find the area between the curve $y = \sin x$, the axis of x and the limits $x = 0$ and $x = \pi$.

4 Find the area of the curve $y^2 = 4x$ cut off by the line $y = x$.

5 Find the area between the curve $y = 4x^2$ and the line $y = 6x - 2$.

6 Find the area between the curve $y = 5x - x^2 - 4$ and the line $y = x - 1$.

7 Find the area in the first quadrant between the straight line $y = 4x$ and the curve $y = x^3$.

In questions 8 to 13, find the volumes of the solids of revolution formed by revolving about the x-axis the areas between the given curves, the x-axis and the given ordinates. Leave your answers in terms of π.

8 $y = \dfrac{x^2}{2}$ $x = 1,\quad x = 3$ 9 $y^2 = x$ $x = 0,\quad x = 2$

10 $y = x + 2$ $x = 1,\quad x = 3$ 11 $xy = 4$ $x = 2,\quad x = 5$

12 $y = x^2 + 2$ $x = -2, x = 2$ 13 $x^2 + y^2 = 9$ $x = -3, x = 3$

In questions 14 to 17, find the volumes formed by rotating the given areas between the curves and the y-axis, about the y-axis.

14 $y = x^4$ $y = 1,\quad y = 4$ 15 $xy = 1$ $y = 1,\quad y = 3$

16 $y = x^2 - 3$ $y = -1, y = 1$ 17 $x^2 + y^2 = a^2$ $y = 0,\quad y = a$

18 Find the volume of a right circular cone of height h and radius of base a. (Take the vertex at the origin and the axis of the cone as the axis of x.)

19 A cone is generated by rotating the line $y = \dfrac{x}{2}$ from $x = 0$ to $x = 10$ about the x-axis. Find its volume.

20 Find the volumes of the solids of revolution formed by rotating about the x-axis the areas between the curves (a) $y = x^2 - 2x$ and the x-axis,
(b) $y = 3x - x^2$ and the x-axis.

21 Find the points of intersection of the curves $3y^2 = x$ and $x^2 = 9y$. Find also the area between them. This area is rotated (a) about the x-axis (b) about the y-axis. Find the volumes of the solids of revolution so formed.

22 Find by integration the volume of a zone of a sphere of radius 8 cm which lies between two parallel planes at 2 cm and 4 cm from the centre and on the same side of it. The equation of a circle of radius r and centre the origin is $x^2 + y^2 = r^2$.

9e The mean value of a function

The mean value of $y = f(x)$ for values of x from a to b is the mean value of the ordinates of the curve for this range.

Let the area under the graph of $y = f(x)$ from $x = a$ to $x = b$ be divided into n strips of equal widths δx and let the mid-ordinates of the strips be $y_1, y_2, y_3, \dots y_n$. See fig. 50.

Figure 50

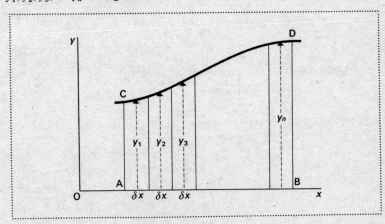

The mean or average value of these ordinates

$$\bar{y} = \frac{y_1 + y_2 + y_3 + \dots + y_n}{n}$$

Multiplying numerator and denominator by δx,

$$\bar{y} = \frac{y_1\,\delta x + y_2\,\delta x + y_3\,\delta x + \ldots + y_n\,\delta x}{n\,\delta x}$$

$$= \frac{\sum y\,\delta x}{b-a} \quad \text{since} \quad n\,\delta x = b-a$$

When the number of strips increase indefinitely, $\delta x \to 0$,

and $\quad \bar{y} = \dfrac{\displaystyle\int_a^b y\,dx}{b-a} = \dfrac{\text{area ABCD}}{b-a}$

In geometric language the mean is the height of the rectangle standing on base AB whose area is equal to the area ABCD beneath the curve.

Example 1 Find the mean value of x^2 between $x = 1$ and 4.

Mean value $\quad \bar{y} = \dfrac{1}{4-1}\displaystyle\int_1^4 x^2\,dx = \dfrac{1}{3}\left[\dfrac{x^3}{3}\right]_1^4 = \dfrac{64-1}{9}$

$$= 7$$

The mean value of a periodic function means its average value over a period unless some other range is specified.

Example 2 Find the mean value of $\cos^2 pt$.

The period of the function $\cos pt$ is $\dfrac{2\pi}{p}$.

Mean value $= \dfrac{1}{\dfrac{2\pi}{p}-0}\displaystyle\int_0^{2\pi/p} \cos^2 pt\,dt$

$$= \dfrac{p}{2\pi}\displaystyle\int_0^{2\pi/p} \dfrac{(1+\cos 2pt)}{2}\,dt$$

$$= \dfrac{p}{4\pi}\left[t + \dfrac{\sin 2pt}{2p}\right]_0^{2\pi/p}$$

$$= \dfrac{p}{4\pi}\left[\dfrac{2\pi}{p} + \dfrac{1}{2p}\sin 4\pi\right]$$

$$= \dfrac{p}{4\pi}\times\dfrac{2\pi}{p} = \dfrac{1}{2}$$

The mean square value of a function

This is the mean value of the second power of the function. Thus the mean square value of y between $x = a$ and $x = b$ is $\dfrac{1}{b-a}\displaystyle\int_a^b y^2\, dx$.

The root mean square (r.m.s.) value of a function

This is the square root of the mean square value

$$\text{r.m.s. value} = \sqrt{\left(\frac{1}{b-a}\int_a^b y^2\, dx\right)}$$

Example 1   Find the r.m.s. value of $i = I \sin pt$ over the range $t = 0$ to $t = \dfrac{2\pi}{p}$.

$$\text{Mean square value} = \frac{1}{\dfrac{2\pi}{p}}\int_0^{2\pi/p} I^2 \sin^2 pt\, dt$$

$$= \frac{pI^2}{2\pi}\int_0^{2\pi/p} \frac{1 - \cos 2pt}{2}\, dt$$

$$= \frac{pI^2}{4\pi}\left[t - \frac{\sin 2pt}{2p}\right]_0^{2\pi/p} = \frac{pI^2}{4\pi}\left[\frac{2\pi}{p} - 0\right]$$

$$= \frac{I^2}{2}$$

Hence r.m.s. value $= \sqrt{\dfrac{I^2}{2}} = \dfrac{I}{\sqrt{2}} = 0{\cdot}7071I$.

Example 2   If $V = V_0 \sin \omega t$ and $C = C_0 \sin(\omega t - \alpha)$ find the mean value of the power product VC.

$$VC = V_0 \sin \omega t\, C_0 \sin(\omega t - \alpha)$$

$$= \frac{V_0 C_0}{2}[\cos \alpha - \cos(2\omega t - \alpha)]$$

The period of VC is $\dfrac{2\pi}{\omega}$.

Hence mean value $= \dfrac{1}{\dfrac{2\pi}{\omega}} \displaystyle\int_0^{2\pi/\omega} \dfrac{V_0 C_0}{2} \left[\cos\alpha - \cos(2\omega t - \alpha) \right] dt$

$$= \frac{\omega V_0 C_0}{4\pi} \left[t \times \cos\alpha - \frac{\sin(2\omega t - \alpha)}{2\omega} \right]_0^{2\pi/\omega}$$

$$= \frac{\omega V_0 C_0}{4\pi} \left[\frac{2\pi}{\omega} \cos\alpha - \frac{\sin(4\pi - \alpha) - \sin(-\alpha)}{2\omega} \right]$$

$$= \frac{\omega V_0 C_0}{4\pi} \times \left(\frac{2\pi}{\omega} \cos\alpha - 0 \right)$$

$$= \tfrac{1}{2} V_0 C_0 \cos\alpha$$

9f Work done by a variable force

Let fig. 51 represent the force–distance diagram of a body moving in a straight line under the action of a force P.

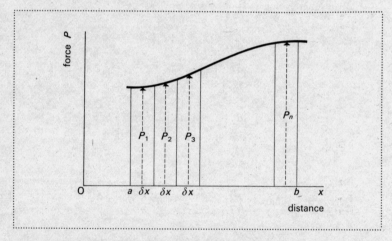

Figure 51

To find the work done by the force when the body moves a distance from $x = a$ to $x = b$ divide the area under the graph into n strips of equal widths δx. Let the mid-ordinates be P_1, P_2, P_3 etc.

The work done in each interval is given approximately by

$P_1 \, \delta x, P_2 \, \delta x, P_3 \, \delta x, \ldots P_n \, \delta x$

Total work done $\simeq P_1 \, \delta x + P_2 \, \delta x + P_3 \, \delta x + \ldots + P_n \, \delta x$

$$\simeq \sum_1^n P \, \delta x$$

The exact amount is the limiting value of this sum as δx approaches 0.

i.e. work done $= \int\limits_a^b P\, dx$

Work done by an expanding gas

When gas expands at a pressure p and moves a piston of area A a distance δx then the work done is given by $pA\,\delta x = p\,\delta v$ where δv is the increase in volume.

The total work done by the gas while it expands from a volume V_1 to V_2 under a variable pressure is given by

Work done $= \int\limits_{V_1}^{V_2} p\, dv$

Example 1 A body moves from rest under the action of a direct force given by $P = \dfrac{20}{x+2}$ N, where x is the distance in metres from the starting point. Find the total work done in moving a distance 6 m, and find the mean value of the force over this distance.

$$\text{Work done} = \int\limits_0^6 P\, dx = \int\limits_0^6 \frac{20}{x+2}\, dx$$

$$= 20\left[\log(x+2)\right]_0^6 = 20\log\frac{8}{2}$$

$$20\log_e 4 = 27\!\cdot\!73\ \text{J} \quad (= \text{N m})$$

$$\text{Mean value of force} = \frac{27\!\cdot\!73\ \text{J}}{6\ \text{m}} = 4\!\cdot\!62\ \text{N}$$

Example 2 A buffer spring is compressed a distance $0\!\cdot\!3$ m. Find the additional work which must be done to compress it a further $0\!\cdot\!2$ m if the spring obeys Hooke's law and its stiffness is such that a force of 2 kN will compress it 1 cm.

Since the spring obeys Hooke's law

Force $P \propto$ compression x

i.e. $\qquad P = Sx$

$$\text{Work done} = \int\limits_{0\cdot3}^{0\cdot5} P\, dx = \int\limits_{0\cdot3}^{0\cdot5} Sx\, dx$$

$$= S\left[\frac{x^2}{2}\right]_{0\cdot3}^{0\cdot5} = \frac{S}{2}(0\!\cdot\!25 - 0\!\cdot\!09)$$

$$= 0\!\cdot\!08S\ \text{m}^2$$

When $x = 0.01$ m, $P = 2$ kN

2 kN $= S\,0.01$ m

$S = 200$ kN m^{-1}

Work done $= 0.08 \times 200$ kN m

$= 16$ kN m

$= 16$ kJ

Example 3 A gas expands according to the law $pv^{1.4} = C$. Initially the pressure is 200 kN m^{-2} when the volume is 1 m^3. Find the work done by the gas in expanding to twice its original volume.

$$pv^{1.4} = C = 200 \times 1^{1.4} \,(\text{kN m})$$

$$C = 200 \text{ and } p = \frac{200}{v^{1.4}}$$

$$\text{Work done} = \int_1^2 p\,dv = \int_1^2 \frac{200}{v^{1.4}}\,dv$$

$$= \frac{200}{-0.4}\left[\frac{1}{v^{0.4}}\right]_1^2$$

$$= 500\left[1 - \frac{1}{2^{0.4}}\right]$$

$$= 500\,[1 - 0.7578]$$

$$= 121 \text{ kJ}$$

Exercise 9f

1 Find the mean value of $x^2 - x$ from $x = 1$ to $x = 3$.

2 Find the average value of $x(3 - x)$ from $x = 0$ to $x = 3$.

3 Find the average value of $\sin x$ between

(a) $x = 0$ and $x = \pi$ (b) $x = 0$ and $x = 2\pi$

4 Find the mean value of (a) $\cos^2 x$ between $x = 0$ and $x = 2\pi$,

(b) $\sin^2 pt$ between $t = 0$ and $t = \dfrac{2\pi}{p}$.

5 Find the mean value of $3\sin 2t$ from $t = 0$ to $t = \dfrac{\pi}{2}$.

6 Find the mean value of $2 + 3\sin x$ from $x = 0$ to $x = \pi$

7 If the velocity v in m s^{-1} of a body is given by $v = 4t+3$ where t is in s, find the average value of v from $t = 2$ to $t = 6$ s.

8 Show that the average value of $y = a+bx+cx^2$ over the range $-h$ to $+h$ is $\frac{1}{6}(y_1+4y_2+y_3)$ where y_1 and y_3 are the first and last ordinates and y_2 is the middle ordinate.

9 A body moves such that the velocity v m s^{-1}, the distance travelled s m and time t s are connected by equations $v = 8t$ and $v^2 = 16s$.
   Find the mean velocity for the first 10 s of motion

   (a) with respect to time   (b) with respect to distance.

10 If $V = 100 \sin \omega t$ and $I = 10 \sin\left(\omega t - \frac{\pi}{3}\right)$, find the mean value of the power VI.

Find the r.m.s. values of the following

11 (a) x from 0 to 2   (b) $x(2-x)$ from $x = 0$ to $x = 2$

12 $1+\sin x$ from $x = 0$ to $x = 2\pi$

13 $3+2 \cos x$ from $x = 0$ to $x = 2\pi$

14 $100 \sin pt+60 \sin 3pt$ from $t = 0$ to $t = \dfrac{2\pi}{p}$

15 The instantaneous potential difference v in an electric circuit is given by $v = 6+8 \cos \omega t$. Find the r.m.s. voltage from $t = 0$ to $\dfrac{2\pi}{\omega}$.

16 The current i milliamps in a circuit is given by $4 \cos \omega t+5 \sin \omega t$. Put i in the form $R \sin (\omega t+\alpha)$ and find the r.m.s. value of the current.

17 (a) A spring of unstretched length 6 cm is extended 3 cm by a pull of 2 N. Find the work done in extending it from 12 cm to 20 cm.
    (b) A spiral spring is stretched a total amount of 50 cm. If the final force is 40 N, find the work done.

18 The couple required to turn a clock spring is given by $C = 0.5+\dfrac{\theta}{20}$ N m where θ is the angle turned through in radians. How much work must be done to wind the spring fully if it takes 12 complete turns?

185 Work done by a variable force

19   The force between two magnetic poles of strengths m_1 and m_2 and placed x cm apart is given by $F = \dfrac{m_1 m_2}{x^2} \times 10^{-5}$ N. Find the work done in moving a pole of strength 5 units from 4 cm to 12 cm away from a pole of strength 6 units.

20   100 m^3 of air at a pressure of 100 kN m^{-2} is compressed until its volume is 40 m^3. Find the amount of work done on the gas if it obeys the law $pv^{1\cdot4} = $ constant.

21   The effective turning moment on the crankshaft of an engine is given by $C = 7800 \sin \theta + 1500 \sin 2\theta$ N m. Find the work done in half a revolution.
      If there are 90 working strokes per minute, calculate the indicated power in kW.

22   A gas expands according to the law $pv = $ constant. When its volume is 4 m^3 the pressure is 300 kN m^{-2}, Find the work done as the gas expands to a volume of 7 m^3.

23   When a gas expands adiabatically, the pressure and volume are connected by the law $pv^\gamma = $ constant. A gas expands from an initial volume v_1 and pressure p_1 to a volume v_2 and pressure p_2. Find the work done by the gas and prove that the average pressure with respect to v during the expansion is $\dfrac{p_1 v_1 - p_2 v_2}{(\gamma - 1)(v_2 - v_1)}$.

9g   Mass and centre of gravity

The mass of a body is given by its volume multiplied by its density. If the density is variable the calculation may involve integration.

Example 1   The mass per unit length of a rod at a distance x m from one end is $\left(2 + \dfrac{x^2}{2}\right)$ kg m^{-1}. If the rod is 5 m long find its total mass.

Figure 52

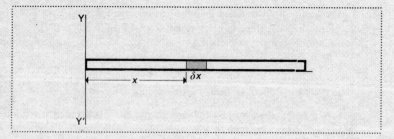

Consider the mass of a small element δx distance x from one end as in fig. 52. For the short length δx it is assumed that the mass per unit length does not change from the value $\left(2+\dfrac{x^2}{2}\right)$ kg m^{-1}.

Mass of element $= \left(2+\dfrac{x^2}{2}\right)\delta x$

Total mass is the sum of the masses of all such elements making up the rod

$$= \sum_{x=0}^{x=5}\left(2+\frac{x^2}{2}\right)\delta x$$

$$= \int_0^5\left(2+\frac{x^2}{2}\right)dx = \left[2x+\frac{x^3}{6}\right]_0^5 = 30\cdot8 \text{ kg}$$

Centre of gravity

Since the resultant weight of a body acts through its centre of gravity the centre of gravity is found by applying the principle of moments, which states that the moment of the resultant of a system of forces about any axis is equal to the sum of the moments of the separate forces.

If bodies of weights w_1, w_2, w_3, etc. are placed in a plane at points (x_1, y_1), (x_2, y_2), (x_3, y_3), etc. respectively then by taking moments about Oy and Ox in turn

$$(w_1+w_2+w_3+\ldots)\,\bar{x} = w_1x_1+w_2x_2+w_3x_3+\ldots$$
$$\text{and} \quad (w_1+w_2+w_3+\ldots)\,\bar{y} = w_1y_1+w_2y_2+w_3y_3+\ldots$$

where $(\bar{x}, \bar{y})$ are the coordinates of the centre of gravity. See fig. 53.

Figure 53

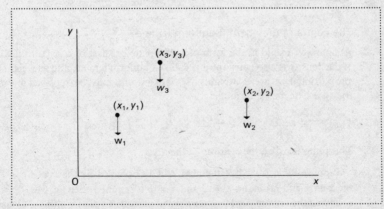

187   **Mass and centre of gravity**

Thus $\quad \bar{x} = \dfrac{\sum wx}{W} \qquad \bar{y} = \dfrac{\sum wy}{W}$

where $W = \sum w =$ total weight.

Example 2 Find the position of the centre of gravity of the rod in example 1.

Let the c.g. be at a distance $\bar{x}$ from YY'.

The weight of the element considered $= \left(2 + \dfrac{x^2}{2}\right) \delta x$

For convenience of taking moments suppose the rod and axis YY' are horizontal.

Moment of weight of element about YY' $= \left(2 + \dfrac{x^2}{2}\right) \delta x \times x$

By the principle of moments

$$(\text{total weight}) \times \bar{x} = \sum \left(2 + \dfrac{x^2}{2}\right) \delta x \times x$$

$$\bar{x} = \frac{\displaystyle\sum_{x=0}^{x=5} \left(2 + \dfrac{x^2}{2}\right) x\, \delta x}{\text{total weight}}$$

$$= \frac{\displaystyle\int_0^5 \left(2x + \dfrac{x^3}{2}\right) dx}{30 \cdot 8} = \frac{\left[x^2 + \dfrac{x^4}{8}\right]_0^5}{30 \cdot 8}$$

$$= 3 \cdot 35$$

Thus the c.g. of the rod is $3 \cdot 35$ m from the lighter end.

The centre of gravity of a uniform lamina

A lamina is a solid in the form of a thin plate or sheet of constant thickness. If the density is also constant it is called a uniform lamina, and as weights are proportional to areas, weights w_1, w_2, w_3, etc. may be replaced by areas a_1, a_2, a_3, etc. in the formulae for c.g.

Thus $\quad \bar{x} = \dfrac{\sum ax}{A} \qquad \bar{y} = \dfrac{\sum ay}{A}$ where $A =$ total area.

This point is called the *centroid* of the area.

Suppose the centroid of the area ABCD bounded by the curve $y = f(x)$, the x-axis, and ordinates $x = a$ and $x = b$ is required. (See fig. 54.)

Divide the area into a large number of narrow strips of width δx. Consider

Figure
54

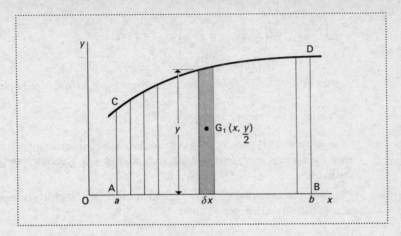

the strip shown shaded of length y and area $y\,\delta x$. The centre of this strip is at its mid point G_1 coordinates $\left(x, \dfrac{y}{2}\right)$.

Then taking moments of area about the y-axis,

$$\bar{x}\text{ (total area of strips)} = \sum_{x=a}^{x=b} (y\,\delta x)\,x$$

$$\bar{x} = \frac{\displaystyle\sum_{x=a}^{x=b} xy\,\delta x}{\displaystyle\sum_{x=a}^{x=b} y\,\delta x} = \frac{\displaystyle\int_a^b xy\,dx}{\displaystyle\int_a^b y\,dx}$$

Similarly by taking moments about the x-axis,

$$\bar{y} = \frac{\displaystyle\sum_{x=a}^{x=b} (y\,\delta x)\,\frac{y}{2}}{\text{total area}} = \frac{\displaystyle\int_a^b \frac{y^2}{2}\,dx}{\displaystyle\int_a^b y\,dx}$$

Example 3   Find the centroid of the area bounded by the curve $y = x^2$ and the ordinates at $x = 1$ and $x = 3$.

$$\bar{x} = \frac{\displaystyle\int_1^3 xy\,dx}{\displaystyle\int_1^3 y\,dx} = \frac{\displaystyle\int_1^3 x^3\,dx}{\displaystyle\int_1^3 x^2\,dx} = \frac{\dfrac{1}{4}\left[x^4\right]_1^3}{\dfrac{1}{3}\left[x^3\right]_1^3} = \frac{3}{4}\frac{(3^4-1)}{(3^3-1)} = 2\tfrac{4}{13}$$

189   Mass and centre of gravity

$$\bar{y} = \frac{\int_1^3 \frac{y^2}{2}\,dx}{\int_1^3 y\,dx} = \frac{\frac{1}{2}\int_1^3 x^4\,dx}{\frac{26}{3}} = \frac{\frac{1}{10}\left[x^5\right]_1^3}{\frac{26}{3}} = 2\frac{103}{130}$$

Centroid is at the point (2·31, 2·79).

Example 4   Find the centre of gravity of the solid of revolution generated when the curve $y = x^2$, from ordinates $x = 0$ to $x = 4$, revolves about the x-axis.

- Let the volume be divided into a large number of discs by planes perpendicular to the x-axis, distance δx apart.

Figure
55

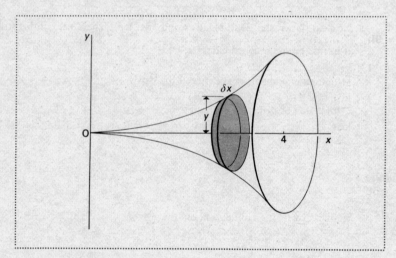

Consider a sample element of radius y, thickness δx as shown in fig. 55. If w is the weight per unit volume

weight of disc $= \pi y^2\,\delta x \times w$

By taking moments about the y-axis of the weights of all the discs, then by the principle of moments

$$(\text{Total weight}) \times \bar{x} = \sum_{x=0}^{x=4} \pi y^2\,\delta x \times wx$$

Since total weight $=$ volume $\times w = w\int_0^4 \pi y^2\,dx$

Thus $\bar{x} = \dfrac{\displaystyle\int_0^4 \pi x y^2 \, dx}{\displaystyle\int_0^4 \pi y^2 \, dx}$

$$= \frac{\pi \displaystyle\int_0^4 x^5 \, dx}{\pi \displaystyle\int_0^4 x^4 \, dx} = \frac{\dfrac{1}{6}\left[x^6\right]_0^4}{\dfrac{1}{5}\left[x^5\right]_0^4} = 3\tfrac{1}{3}$$

By symmetry $\bar{y} = 0$.

C.G. of solid of revolution is at point $(3\tfrac{1}{3}, 0)$.

9h   Theorems of Pappus (Guldin)

I   When a plane curve revolves about an axis in its plane, not intersecting it, the area of the surface generated is equal to the length of the arc multiplied by the length of the path traced out by the centroid of the curve.

Let s be the total length of the curve. Divide the curve into small elements of length δs and suppose the centroid of a sample element is at a distance y from the axis. (See fig. 56.)

Figure 56

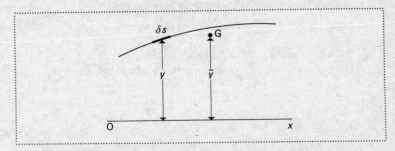

The area of the surface swept out by δs when it makes a complete revolution is approximately $2\pi y \, \delta s$.

As δs is made smaller and smaller the more exact this is.

Total area of surface generated $= \sum 2\pi y \, \delta s$

But $\qquad \sum y \times \delta s = s\bar{y}$

Area of surface $= 2\pi s \bar{y} = s \times 2\pi \bar{y}$

$\qquad\qquad\quad$ = length of curve × path traced out by c.g.

The rotation has been taken for a complete revolution but the formula obviously holds for any part of a complete revolution.

II When a plane area revolves about an axis in its plane not cutting the area, the volume generated is equal to the area multiplied by the length of the path traced out by the centroid of the area.

Consider a small element δA at a distance y from the axis. (See fig. 57.)

Figure
57

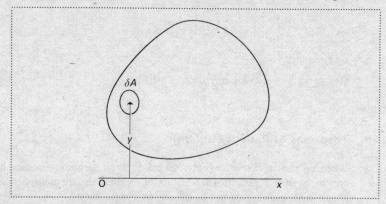

Volume of ring traced out by δA in one revolution $\simeq 2\pi y\,\delta A$.

Volume of whole solid $\simeq \sum 2\pi y\,\delta A$, the approximation becoming more and more accurate the smaller δA is made.

If $\bar{y}$ is the distance of the centroid from the axis

$$A\bar{y} = \left(\sum \delta A\right)\bar{y} = \sum y\,\delta A \quad \text{by the principle of moments.}$$

Thus volume of solid $= 2\pi A\bar{y}$
$\qquad\qquad\qquad\qquad = \text{area} \times \text{path traced out by centroid.}$

Figure
58

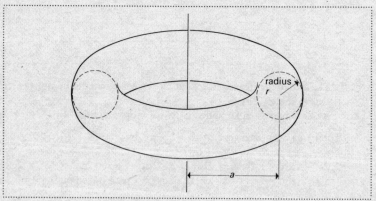

radius
r

a

192 Applications of integration

Example 1 Find the area of the surface and the volume of an anchor ring formed by the rotation of a circle of radius r about an axis distance a from the centre. See fig. 58.

Area of surface $= 2\pi r \times 2\pi a$
$\qquad\qquad\quad = 4\pi^2 ar$
Volume of ring $= \pi r^2 \times 2\pi a$
$\qquad\qquad\quad = 2\pi^2 ar^2$

The theorems may be used, conversely, to find the centroid of a plane arc or a plane area when the surface or volume generated by revolution is known independently.

Example 2 Find the centroids of a semicircular arc and a semicircular area each of radius r. See fig. 59.

Figure 59

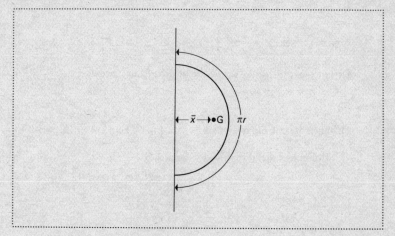

i Area swept out by the revolution of a semicircular arc is the surface of a sphere:

$$4\pi r^2 = \pi r \times 2\pi \bar{x}_1 \quad \text{by Pappus}$$

Hence $\bar{x}_1 = \dfrac{4\pi r^2}{2\pi^2 r} = \dfrac{2r}{\pi}$ (for the arc)

ii Volume swept out by the revolution of a semicircular area is that of a solid sphere:

$$\frac{4}{3}\pi r^3 = \frac{\pi r^2}{2} \times 2\pi \bar{x}_2 \quad \text{by Pappus}$$

Hence $\bar{x}_2 = \dfrac{4\pi r^3}{3\pi^2 r^2} = \dfrac{4r}{3\pi}$ (for the area)

Example 3   A steel shaft 8 cm diameter has a semicircular groove turned out of it to a depth of $1\frac{1}{2}$ cm. Calculate the volume of metal removed. See fig. 60.

Figure 60

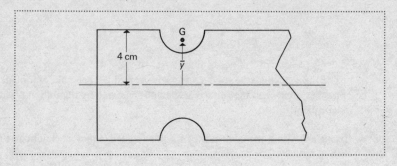

Area of semicircle $= \dfrac{1}{2} \pi \left(\dfrac{3}{2}\right)^2 = \dfrac{9}{8} \pi \text{ cm}^2$

Distance of centroid of semicircular area from its centre $= \dfrac{4r}{3\pi} = \dfrac{4(\frac{3}{2})}{3\pi}$

$$= \dfrac{2}{\pi} \text{ cm}$$

Distance of centroid from axis $= \bar{y} = \left(4 - \dfrac{2}{\pi}\right) \text{ cm}$

Volume of metal removed $= \text{area} \times 2\pi\bar{y}$

$$= \dfrac{9\pi}{8} \times 2\pi \left(4 - \dfrac{2}{\pi}\right) = 9\pi^2 - \dfrac{9}{2}\pi$$

$$= 88\!\cdot\!8 - 14\!\cdot\!1$$

$$= 74\!\cdot\!7 \text{ cm}^3$$

Exercise 9h

1   The mass per unit length of a billiard cue is $(1\!\cdot\!2 + 0\!\cdot\!02x) \text{ g cm}^{-1}$ where x is the distance from the tip in centimetres. Find its mass and the distance of its centre of gravity from the tip if its length is $1\!\cdot\!5$ m.

2   The mass per unit length of a straight rod at x m from one end is $\left(1 + \dfrac{x}{2}\right) \text{kg m}^{-1}$.

If the rod is 6 m long find its mass and the distance of its c.g. from the lighter end.

3   A conical mound of earth is 6 m high and has a base radius of 10 m. Due to the greater compression near the base the density varies and is $(1600 + 2x) \text{ kg m}^{-3}$ at x m below the vertex. Find its mass.

194   Applications of integration

In examples 4, 5, 6, 7, find the positions of the centroids of the uniform laminae with the given boundaries.

4   $y = x^2$; the x-axis, $x = 2$.

5   $y = 2x^3$; the x-axis, $x = 1, x = 2$.

6   The part of $y = 4 - x^2$ above the x-axis.

7   The portion of $y = 2x - x^2$ above the x-axis.

In examples 8, 9, 10, 11, find the positions of the c.g.s of the solids of revolution generated by the curves about the axes stated.

8   $y = x^2$ from $x = 0$ to $x = 2$ about the x-axis.

9   $y = x^2$ from $x = 0$ to $x = 2$ about the y-axis.

10   $y = x^2 - x^3$ from $x = 0$ to $x = 1$ about the x-axis.

11   The cone obtained by rotating $y = \dfrac{r}{h}x$ from $x = 0$ to $x = h$ about the x-axis.

12   Find the centroid of the area bounded by the curves $y^2 = 4x$ and $x^2 = 4y$.

13   Find the centroid of the area under the curve $y = \sin 2x$ from $x = 0$ to $\dfrac{\pi}{2}$.

14   Find the c.g. of the hemisphere of uniform density formed by rotating the first quadrant of the circle $x^2 + y^2 = a^2$ about the x-axis.

15   Find the c.g. of the hemisphere formed by rotating the first quadrant of the circle $x^2 + y^2 = a^2$ about the x-axis when the density at any point is directly proportional to its distance from the base, i.e. density $= kx$.

16   A ring has an external diameter of 3 cm and its cross-section is a circle of diameter 0·5 cm. Find the area of the surface and the volume of the ring.

17   A semicircular bend of lead pipe has a mean radius of 12 cm. The internal diameter of the pipe is 4 cm and the thickness of the walls is 0·5 cm. Find its mass given the density of lead is 11·3 g cm^{-3}.

18   The radial section of the rim of a wheel is an isosceles triangle of base 1·5 cm and height 2 cm. If the outer diameter of the wheel is 10 cm find the mass of the rim when its material has a density of 2·56 Mg m^{-3}.

195   Theorems of Pappus (Guldin)

19　A bevelled ring is formed by rotating an isosceles triangle of base 0.5 cm and equal sides 1 cm about an axis parallel to the base and distant 2 cm below it.

　　Find (a) the volume, (b) the mass of the ring whose material has a density of 7.7 g cm^{-3}.

20　A solid steel disc, 8 cm diameter and 2 cm thick, has a semicircular groove of 1.5 cm diameter sunk into its rim so as to form a pulley wheel. Find its mass if the density is 7.8 g cm^{-3}.

21　Find the centroid of the semi-annulus bounded by two concentric semi-circles of radii r and R $(r < R)$. State and interpret the result for the special cases where (a) $r = 0$, (b) $r = R$.

22　A 12-cm diameter shaft has an isosceles triangular groove 8 cm wide and 1.5 cm deep turned out of it in a lathe. Calculate the volume of metal removed, both by the theorem of Pappus and by considering the volumes of the conical parts which remain.

9i　Moments of inertia

The moment of inertia of a set of particles about an axis is the sum of the products of the mass of each particle and the square of its distance from the axis.

　　Denoting this by I we have

$$I = m_1 r_1^2 + m_2 r_2^2 + m_3 r_3^2 + \ldots$$
　　where $m_1, m_2, m_3, \ldots$ are the masses of the particles
　　and $r_1, r_2, r_3, \ldots$ their respective distances from the axis.
$$= \sum mr^2$$

It may be regarded as the *second moment of mass* about the given axis. For continuous bodies where the number of particles is infinitely great, the summation is replaced by an integral.

　　If M is the total mass of the set of particles and k the distance from the axis at which a particle of mass M would have the same moment of inertia as the particles then

$$Mk^2 = I = \sum mr^2$$
$$\text{and} \quad k^2 = \frac{I}{M} = \frac{\sum mr^2}{M}$$

k is known as the *radius of gyration* of the particles about the axis.

Second moment of area

If δA is an element of an area at a perpendicular distance x from an axis in its plane, then $x^2 \, \delta A$ is called the second moment of area about the axis. The sum for all the elements of area $\sum x^2 \, \delta A$ is the second moment of the whole area and it may be denoted by Ak^2

whence $\quad k^2 = \dfrac{\sum x^2 \, \delta A}{A}$

The moment of inertia is needed when dealing with the rotation of a solid body about an axis. One may think of it as playing the same role in rotational motion as mass does in linear motion. The formula $\frac{1}{2}mV^2$ for kinetic energy of a body moving in a straight line becomes $\frac{1}{2}I\omega^2$ for the kinetic energy of a body rotating about an axis.

Second moments of area occur in problems relating to bending of beams and torsion of shafts, where they are an indication of the stiffness of the members.

1 Moment of inertia of a uniform rod

(a) About an axis through its centre normal to its length.

Let m be the mass per unit length of the rod of length l.

Consider a small element of length δx at a distance x from axis YY'. See fig. 61.

Figure 61

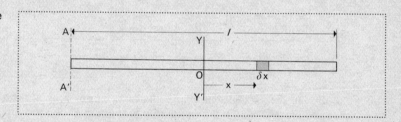

Mass of element $= m \, \delta x$

M.I. of element about YY' $= (m \, \delta x)x^2 = mx^2 \, \delta x$

$$\text{M.I. of rod } I_{YY'} = \sum_{x=-\frac{1}{2}l}^{x=+\frac{1}{2}l} mx^2 \, \delta x$$

$$= \int_{-\frac{1}{2}l}^{+\frac{1}{2}l} mx^2 \, dx = m \left[\frac{x^3}{3} \right]_{-\frac{1}{2}l}^{+\frac{1}{2}l}$$

Since the total mass of the rod $M = ml$

then $\quad I_{YY'} = \dfrac{Ml^2}{12} \quad$ and $\quad k^2_{YY'} = \dfrac{l^2}{12}$

(b) About a perpendicular axis at one end.

In this case the distance x is measured from the axis AA' and the limits of integration become 0 to l.

$$I_{AA'} = \int\limits_0^l mx^2 \, dx = \frac{ml^3}{3}$$

$$= \frac{Ml^2}{3} \quad \text{and} \quad k_{AA'}^2 = \frac{l^2}{3}$$

2   Second moment of area of a rectangle

The axis YY′ is in the plane of the rectangle and bisects a pair of opposite sides. See fig. 62.

Figure 62

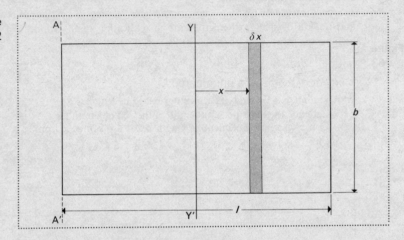

By dividing the rectangle into a large number of narrow strips of width δx, the second moment of area of a typical strip becomes $x^2 \, \delta A = x^2 b \, \delta x$

$$\text{Second moment of area} = \sum_{x=-\frac{1}{2}l}^{x=\frac{1}{2}l} bx^2 \, \delta x = \int\limits_{-\frac{1}{2}l}^{\frac{1}{2}l} bx^2 \, dx$$

$$= b \left[\frac{x^3}{3} \right]_{-\frac{1}{2}l}^{\frac{1}{2}l} = \frac{l^3 b}{12}$$

Since   total area $A = lb$ then

$$\text{Second moment of area} = \frac{Al^2}{12} = Ak_{YY'}^2$$

$$k_{YY'}^2 = \frac{l^2}{12}$$

If the axis is in one side of the rectangle, say AA′ in fig. 62, then the second moment of area is $\frac{Al^2}{3}$ and $k_{AA'}^2 = \frac{l^2}{3}$

If a rectangular lamina has a mass M then its m.i. about a side will be $Mk^2 = \dfrac{Ml^2}{3}$.

Thus the m.i. of a uniform rectangular metal door of mass 100 kg and 2 m high by 1 m wide about its hinges $= Mk^2 = \dfrac{100 \times 1^2}{3}$ kg m^2 $= 33.3$ kg m^2.

3   M.I. of a circular disc about an axis through its centre perpendicular to its plane

Divide the disc into thin concentric rings of width δx. See fig. 63.

Figure 63

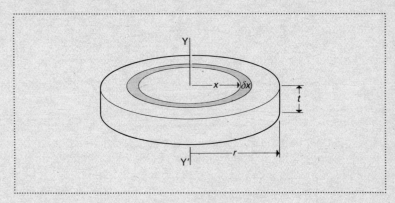

All the material inside a ring of radius x, width δx, and depth t will be approximately at the same distance x from the axis.

$$\text{M.I. of ring} = \text{mass} \times x^2 = (2\pi x\, \delta x \times t\rho)\, x^2$$
$$\text{where } \rho = \text{density of material.}$$
$$\text{M.I. of disc} = \sum_{x=0}^{x=r} 2\pi t\rho x^3\, \delta x$$

$$= 2\pi t\rho \int_0^r x^3\, dx = \frac{2\pi t\rho r^4}{4}$$

Since mass of disc $M = \pi r^2 t\rho$

$$\text{m.i.}_{YY'} = M\frac{r^2}{2} \quad \text{and} \quad k_{YY'}^2 = \frac{r^2}{2}$$

Calculation of second moments of mass and of area is very much simplified by the aid of the following theorems.

1 The perpendicular axes theorem

If OX, OY, OZ are three mutually perpendicular axes and OX and OY lie in the plane of a lamina, then

$$I_{OZ} = I_{OX} + I_{OY} \quad \text{and} \quad k_{OZ}^2 = k_{OX}^2 + k_{OY}^2$$

where I_{OX} is the m.i. about axis OX
and k_{OX} is radius of gyration about OX, etc.

Let the lamina, lying in the plane of XOY, be divided into small particles of mass δm. See fig. 64.

Figure 64

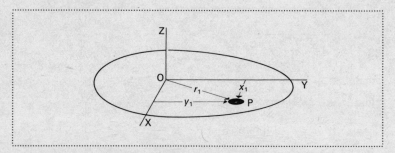

Let δm_1 be the mass of a particle at P (x_1, y_1) and
δm_2 be the mass of a particle at (x_2, y_2), etc.

Then
$$I_{OX} = \delta m_1 \, y_1^2 + \delta m_2 \, y_2^2 + \dots$$
$$I_{OY} = \delta m_1 \, x_1^2 + \delta m_2 \, x_2^2 + \dots$$
$$I_{OX} + I_{OY} = \delta m_1 \, (x_1^2 + y_1^2) + \delta m_2 \, (x_2^2 + y_2^2) + \dots$$
$$= \delta m_1 \, r_1^2 + \delta m_2 \, r_2^2 + \dots$$

where $r_1, r_2, \dots$ are the distances from axis OZ.

$$I_{OX} + I_{OY} = I_{OZ}$$

Dividing both sides by the total mass M

$$k_{OX}^2 + k_{OY}^2 = k_{OZ}^2$$

2 The parallel axes theorem

If GZ is an axis through the centre of gravity of a body and HH' is a parallel axis distance h from it, then

$$I_{HH'} = I_{GZ} + Mh^2 \quad \text{and} \quad k_{HH'}^2 = k_{GZ}^2 + h^2$$

In fig. 65 let GZ be the axis through the c.g. of the body. Take GX, GY and GZ as three mutually perpendicular axes and let HH' be an axis parallel to GZ and distance h from it and lying in the plane XGZ.

Figure
65

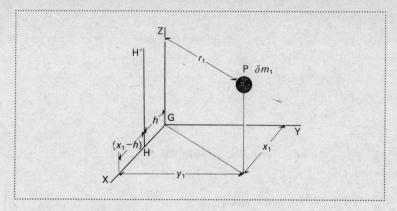

Let P be a particle of mass δm_1 at distances x_1 and y_1 from planes YGZ and XGZ respectively.

(distance)^2 of P from axis HH′ $= (x_1 - h)^2 + y_1^2$

$$= x_1^2 + y_1^2 + h^2 - 2hx_1$$

$$= r_1^2 + h^2 - 2hx_1$$

where $r_1 =$ its distance from GZ

Let Q be a particle of mass δm_2 at distances x_2, y_2 and R a particle mass δm_3 at distances x_3, y_3, etc. Then m.i. of body about HH′ is given by

$$I_{HH'} = \delta m_1 (r_1^2 + h^2 - 2hx_1) + \delta m_2 (r_2^2 + h^2 - 2hx_2) + \delta m_3 (r_3^2 + h^2 - 2hx_3) + \ldots$$

$$= (\delta m_1 r_1^2 + \delta m_2 r_2^2 + \ldots) + h^2(\delta m_1 + \delta m_2 + \ldots) - -2h(\delta m_1 x_1 + \delta m_2 x_2 + \ldots)$$

$$= I_{GZ} + Mh^2 - 2h \sum x \, \delta m$$

Since GZ goes through the c.g. of the body $\bar{x} = 0$

$$\bar{x} = \frac{\sum x \, \delta m}{M} = 0$$

Thus $I_{HH'} = I_{GZ} + Mh^2$

and dividing by the mass M

$$k_{HH}^2 = k_{GZ}^2 + h^2$$

3 The m.i. of a circular disc
(a) about a diameter, (b) about a tangent

(a) Let the disc be placed as in fig. 66 with axes OX and OY in the plane and OZ normal to the plane of the disc through the centre.

Figure
66

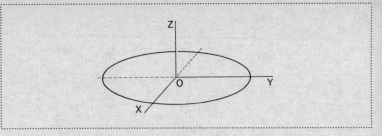

By the perpendicular axis theorem

$$I_{OZ} = I_{OX} + I_{OY}$$

$I_{OX} = I_{OY}$ by symmetry, and $I_{OZ} = \dfrac{Mr^2}{2}$

$$I_{OX} = I_{OY} = \frac{Mr^2}{4}$$

Figure
67

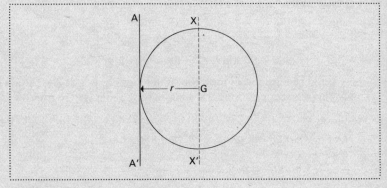

(b) By the parallel axis theorem

$$I_{AA'} = I_{XX'} + Mr^2 \quad \text{(see fig. 67)}$$
$$= \frac{Mr^2}{4} + Mr^2$$
$$= \frac{5Mr^2}{4}$$

4 Second moment of an area bounded by a curve, the x-axis and ordinates $x = a$, $x = b$

Divide the area into a large number of narrow strips of width δx as in fig. 68.

If δA is the area of a typical strip, $\delta A = y\,dx$ and since k^2 for this element is

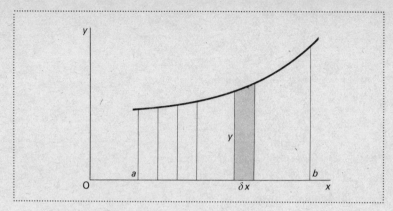

Figure 68

(a) x^2 about Oy (b) $\dfrac{y^2}{3}$ about Ox

then $Ak_{Oy}^2 = \sum x^2 y\,\delta x$

and $Ak_{Ox}^2 = \sum \dfrac{y^2}{3}\, y\,\delta x$

Thus $k_{Oy}^2 = \dfrac{\displaystyle\int_a^b x^2 y\,dx}{\displaystyle\int_a^b y\,dx}$ and $k_{Ox}^2 = \dfrac{\displaystyle\int_a^b \dfrac{1}{3}\,y^3\,dx}{\displaystyle\int_a^b y\,dx}$

Example Find the radius of gyration about both Ox and Oy of the area in the first quadrant, between the curve $y^2 = 4x$, the x-axis and the ordinate $x = 4$. See fig. 69.

Figure 69

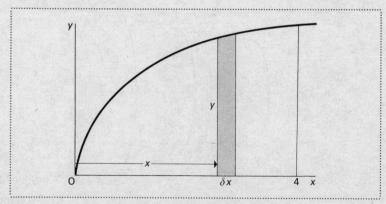

i About the y-axis

$$k_{Oy}^2 = \dfrac{\displaystyle\int_0^4 x^2 y\, dx}{\displaystyle\int_0^4 y\, dx} = \dfrac{\displaystyle\int_0^4 2x^{\frac{5}{2}}\, dx}{\displaystyle\int_0^4 2x^{\frac{1}{2}}\, dx}$$

$$= \dfrac{\dfrac{4}{7}\times 2^7}{\dfrac{4}{3}\times 2^3} = 6\tfrac{6}{7}$$

$$k_{Oy} = 2{\cdot}62$$

(ii) About the x axis

$$k_{Ox}^2 = \dfrac{\displaystyle\int_0^4 \tfrac{1}{3} y^3\, dx}{\displaystyle\int_0^4 y\, dx} = \dfrac{\displaystyle\int_0^4 \tfrac{8}{3} x^{\frac{3}{2}}\, dx}{\dfrac{4}{3}\times 2^3} = \dfrac{\dfrac{8}{3}\times\dfrac{2}{5}\times 2^5}{\dfrac{4}{3}\times 2^3} = \dfrac{16}{5}$$

$$k_{Ox} = 1{\cdot}79$$

The m.i. of a solid of revolution

Divide the solid into a large number of thin discs, thickness δx, and perpendicular to the axis of revolution, the x-axis in this case.

Figure 70

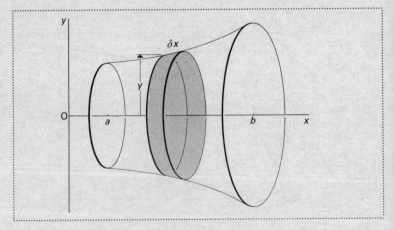

Let y be the radius of a typical disc distance x from O. (See fig. 70.)
Then the mass of this element $\delta m = \rho\pi y^2\, \delta x$ where $\rho = $ density.

The m.i. of this element about $\text{O}x = \delta m \dfrac{y^2}{2} = \rho \dfrac{\pi y^4}{2}\, \delta x$

Total m.i. about $\text{O}x = I_{\text{O}x} = \displaystyle\sum \rho \dfrac{\pi y^4}{2}\, \delta x = \int_a^b \dfrac{\rho\pi}{2}\, y^4\, dx$

$$= \text{Mass} \times k_{\text{O}x}^2 = k_{\text{O}x}^2 \int_a^b \rho\pi y^2\, dx$$

$$k_{\text{O}x}^2 = \dfrac{\dfrac{\rho\pi}{2} \displaystyle\int_a^b y^4\, dx}{\rho\pi \displaystyle\int_a^b y^2\, dx} = \dfrac{\dfrac{1}{2} \displaystyle\int_a^b y^4\, dx}{\displaystyle\int_a^b y^2\, dx}$$

The m.i. of this same element about Oy is, by the parallel axis theorem,
$\delta m \left(\dfrac{y^2}{4} + x^2 \right)$.

$$\text{Hence} \quad k_{\text{O}y}^2 = \dfrac{\displaystyle\int_a^b \rho\pi y^2 \left(\dfrac{y^2}{4} + x^2 \right) dx}{\displaystyle\int_a^b \rho\pi y^2\, dx}$$

Example Find the m.i. of a right circular cone about its axis.

Let the cone be placed as in fig. 71 with its axis along the x-axis.
Treating it as a solid of revolution the equation of the line generating it is
given by $\dfrac{y}{x} = \dfrac{r}{h}$

i.e. $\quad y = \dfrac{r}{h} x$

Mass $\quad M = \frac{1}{3}\pi r^2 h \rho$

$$\text{m.i.} = \int_0^h \dfrac{\rho\pi}{2} y^4\, dx = \dfrac{\rho\pi}{2} \int_0^h \dfrac{r^4}{h^4} x^4\, dx = \dfrac{\rho\pi}{2} \dfrac{r^4}{h^4} \left[\dfrac{h^5}{5} \right]$$

$$= \dfrac{\rho\pi r^4 h}{10} = \dfrac{\rho\pi r^2 h}{3} \times \dfrac{3}{10} r^2 = M \dfrac{3}{10} r^2$$

Figure 71

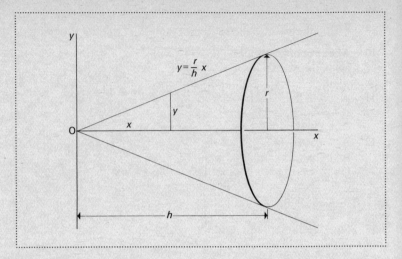

$y = \dfrac{r}{h} x$

Exercise 9i

1   Calculate the radius of gyration of a rectangular door 2 m high by 1·2 m wide about a vertical axis through its hinges.

2   A circular door of a boiler is hinged so that it turns about a tangent. If its diameter is 0·8 m and its mass 50 kg, find its m.i. and radius of gyration about the hinge.

3   Find the radius of gyration of a square of side l about a diagonal.

4   The mass per unit length of a straight rod at x m from one end is $\left(1 + \dfrac{x}{2}\right)$ kg m^{-1}. If the rod is 6 m long, find k^2 about the lighter end.

5   The mass per unit length of a billiard cue is $(1·2 + 0·02\,x)$ g cm^{-1} where x is the distance from the tip. Find k^2 about the tip if the cue is 1·5 m long.

6   Find the second moment of area of the figure formed by the curve $y = x^2$, the x-axis and the ordinate $x = 2$, (a) about y-axis   (b) about x-axis.
     Find also k_{Oy}^2 and k_{Ox}^2.

7   A plane area consists of the portion of the curve $y = 4 - x^2$ above the x-axis. Find k_{Ox}^2 and k_{Oy}^2.

8   Find k_{Ox}^2 and k_{Oy}^2 for the portion of the curve $y = x - x^2$ lying above the x-axis.

9   A solid is formed by the revolution of the curve $y = x^2$ from $x = 0$ to $x = 2$ about the x-axis. Find k_{Ox}^2 and k_{Oy}^2 for this solid.

10   A solid of revolution is formed by rotating $y = x^2 - x^3$ from $x = 0$ to $x = 1$ about the x-axis. Find k_{Ox}^2 and k_{Oy}^2.

11   A cone is formed by rotating the line $y = x \tan \alpha$ from $x = 0$ to $x = h$ about the x-axis. Find the m.i. about its axis.

12   A triangle is of height h and base b. By dividing it into elemental strips parallel to the base and distance x from the vertex, find:
(a) k^2 about an axis through the vertex parallel to the base,
(b) k^2 about an axis through its c.g. parallel to the base,
(c) k^2 about the base.

13   Show by integration that the radius of gyration of a cylinder of height h and radius r about an end diameter is $\sqrt{\left(\dfrac{r^2}{4} + \dfrac{h^2}{3}\right)}$.

14   Show that the second moment of area of a rectangle sides b and d about an axis coinciding with side b is $\dfrac{bd^3}{3}$.

     Using this formula, find the second moment of area about XX' of the beam section shown in fig. 72 by considering half the section as rectangle ABCD $- 2 \times$ rectangle DEFG.

Figure 72

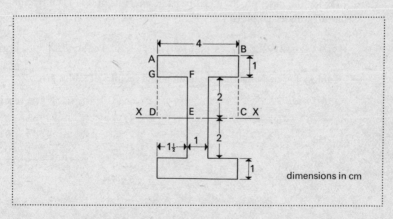

dimensions in cm

15   Find the second moment of area of the beam section in fig. 73 about axis XX' and about a parallel axis through the centroid.

Figure
73

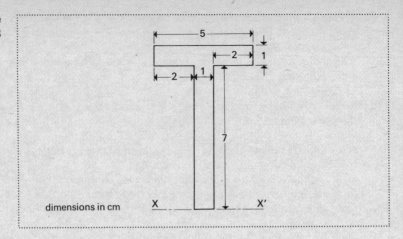

dimensions in cm

9j Approximate evaluation of definite integrals

When an indefinite integral is not known, an approximate value for the definite integral may be obtained by finding the area under its graph.

Any of the rules, mid-ordinate, trapezoidal, or Simpson's may be used to find the area.

Example 1 Find the value of $\int_0^1 e^{-x^2}\, dx$.

Divide the base into 10 equal parts at intervals of $0\cdot1$. Then by Simpson's rule, area under graph is

$$\frac{h}{3}\left[y_1 + y_{11} + 4(y_2 + y_4 + \ldots + y_{10}) + 2(y_3 + y_5 + \ldots + y_9)\right]$$

Tables of e^x or Napierian logarithms and reciprocal tables give

$$
\begin{aligned}
y_1 &= e^0 = 1 \\
y_{11} &= e^{-1} = 0\cdot3679 \\
\hline
y_1 + y_{11} &= 1\cdot3679
\end{aligned}
$$

$$
\begin{aligned}
y_2 &= e^{-0\cdot01} = 0\cdot9900 \\
y_4 &= e^{-0\cdot09} = 0\cdot9139 \\
y_6 &= e^{-0\cdot25} = 0\cdot7788 \\
y_8 &= e^{-0\cdot49} = 0\cdot6126 \\
y_{10} &= e^{-0\cdot81} = 0\cdot4449 \\
\hline
y_2 + \ldots + y_{10} &= 3\cdot7402
\end{aligned}
$$

$$
\begin{aligned}
y_3 &= e^{-0\cdot04} = 0\cdot9608 \\
y_5 &= e^{-0\cdot16} = 0\cdot8521 \\
y_7 &= e^{-0\cdot36} = 0\cdot6977 \\
y_9 &= e^{-0\cdot64} = 0\cdot5273 \\
\hline
y_3 + \ldots + y_9 &= 3\cdot0379
\end{aligned}
$$

Thus $\int_0^1 e^{-x^2}\, dx \simeq \dfrac{0\cdot1}{3}\left[1\cdot3679 + 4(3\cdot7402) + 2(3\cdot0379)\right]$

$$\simeq 0\cdot747$$

Example 2 Evaluate $\int_1^{16} \sqrt{x}\, dx$ by the trapezoidal rule using 5 strips. Check your answer by direct integration.

x	1	4	7	10	13	16
$y = \sqrt{x}$	1	2	2·649	3·162	3·606	4

$$\text{Area} = h\left(\frac{y_1+y_6}{2}+y_2+y_3+y_4+y_5\right)$$

$$= 3\left(\frac{1+4}{2}+2+2\cdot649+3\cdot162+3\cdot604\right) = 3 \times 13\cdot917$$

$$= 41\cdot8$$

$Check \quad \int_1^{16} \sqrt{x}\, dx = \left[\frac{2}{3}x^{\frac{3}{2}}\right]_1^{16} = \frac{2}{3}(64-1) = 42$

Exercise 9j

1 Using Napierian logarithmic tables, find $\log_e 3$, $\log_e 3\cdot5$, $\log_e 4$ to two decimal places. Hence using Simpson's rule find approximately $\int_3^4 \log_e x\, dx$.

2 Find the value of $\int_1^6 \dfrac{dx}{5+x}$ using the following table of values

x	1	2	3	4	5	6
$\dfrac{1}{5+x}$	0·167	0·143	0·125	0·111	0·100	0·091

In the following examples evaluate the integrals by Simpson's rule, using the number of strips indicated.

3 $\int_4^{16} \sqrt{x}\, dx$ 6 strips   4 $\int_1^{2\cdot6} \dfrac{dx}{x}$ 8 strips

5 $\int_1^{2\cdot2} \log_{10} x\, dx$ 6 strips   6 $\int_0^1 10^x\, dx$ 10 strips

7 $\int_0^{\frac{1}{2}\pi} \sqrt{(\cos\theta)}\, d\theta$ 4 strips   8 $\int_0^3 \sqrt{(3x+1)}\, dx$ 6 strips

9 $\int_1^2 \sqrt{(1+x^2)}\,dx$ 10 strips   10 $\int_1^5 \dfrac{dx}{2+x}$ 8 strips

Miscellaneous exercises 9

1 (a) Sketch the part of the curve $y = (x-2)(3-x)$ between $x = 0$ and $x = 3$ and find the coordinates of its turning point.
(b) Find the area enclosed by the curve and the axis of x.
(c) Find the distance of the centroid of this area from the axis of x.

N.C.T.E.C.

2 (a) Sketch (do not plot on squared paper) the curves $y^2 = 4x$ and $x^2 = 4y$ showing clearly the area enclosed between them.
(b) Find this area and the volume of the solid formed by rotating it through $360°$ about the axis of x.
(c) Deduce the distance of the centroid of this area from the axis of x.

N.C.T.E.C.

3 Sketch the curve $y = 3x - x^2$ from $x = 0$ to $x = 3$.
    The area under this curve from $x = 0$ to $x = 3$ is rotated through $360°$ about the x-axis generating a volume V_x and then rotated through $360°$ about the y-axis generating a volume V_y. Show that $\dfrac{V_x}{V_y} = \dfrac{3}{5}$.

U.E.I.

4 A uniform solid is formed by the revolution about the y-axis of that portion of the curve $y = \dfrac{x^2}{a}$ which lies between the origin and the point (a, a).
    Sketch the solid so formed and find its volume and the position of its centre of gravity.

D.T.C.

5 (a) Sketch the curves: i $y = \dfrac{x^2}{4}$, ii $y = 8 - 2\sqrt{x}$, from $x = 0$ to $x = 4$.
(b) Find the area in the first quadrant enclosed by the curves

$$y = \dfrac{x^2}{4}, y = 8 - 2\sqrt{x} \text{ and the } y\text{-axis.}$$

(c) Find the y-coordinate of the centroid of this area.

U.E.I.

6   Sketch the curves $y = x^2$ and $y = 6 - \frac{1}{2}x^2$ from $x = -2$ to $x = +2$.

Calculate the area enclosed by the curves and the volume generated when this area is revolved through $360°$ about the axis of x.

Deduce the coordinates of the centroid of the area.

E.M.E.U.

7   Sketch, on the same diagram with the same axes, the curves

$$y = 4x - 2x^2 \quad \text{and} \quad y = 4 - 4x + x^2$$

Calculate the coordinates of the points where the two curves intersect, and find the area included between the two curves.

D.T.C.

8   Sketch, approximately to scale, the curve $y^2 = 9x$ between $x = 0$ and $x = 9$.

Find the coordinates of the centroid of the area above the x-axis bounded by the curve and the ordinate $x = 9$. This area is rotated through $360°$ about the x-axis.

Find the volume swept out.

U.L.C.I.

9   Sketch the curve $y = (1+x)(3-x)$ from $x = -1$ to $x = 3$.

Show that the area S enclosed by the curve and the x-axis is divided in the ratio $5:27$ by the y-axis.

Write down the x-coordinate of the centroid of the area S and calculate the x-coordinate of the centroid of that part of S that lies in the first quadrant. Deduce the x-coordinate of the centroid of the other part of S.

E.M.E.U.

10   (a) Evaluate: $\displaystyle\int_1^3 \left(x^{\frac{1}{2}} + x^{-\frac{1}{2}} \right)^2 dx$ and $\displaystyle\int_0^3 \frac{x+2}{x^2 + 4x + 4} dx$

(b) Show that the ratio of the r.m.s. value to the mean value of the function

$y = \sin x$ over the interval $0 \leqslant x \leqslant \pi$ is $\dfrac{\pi}{2\sqrt{2}}$.

Note: mean value of function $f(x)$, $a \leqslant x \leqslant b$, is $\dfrac{1}{b-a} \displaystyle\int_a^b f(x)\, dx$.

R.M.S. value of function $f(x)$, $a \leqslant x \leqslant b$, is $\sqrt{\left(\dfrac{1}{b-a} \displaystyle\int_a^b [f(x)]^2 dx \right)}$.

U.E.I.

11   (a) Express $\cos 4x \cos x$ as the sum of two cosines and hence evaluate

$$\int_0^{\frac{1}{4}\pi} \cos 4x \cos x\, dx$$

211   **Miscellaneous exercises 9**

(b) i State (without proof) Simpson's rule for approximate integration.

ii The values of a voltage over a period at intervals of 0·01 second are

0, 9, 16, 21, 24, 25, 24, 21, 16, 9, 0

Find, by Simpson's rule, the r.m.s. value of the voltage over the period.

D.T.C.

12 (a) Evaluate: i $\int_1^4 3x^{-2\cdot5}\,dx$, ii $\int_{-1}^1 e^{2x}\,dx$

(b) The pressure p and volume v of a given mass of gas are connected by the relation

$$\left(p+\frac{a}{v^2}\right)(v-b) = k$$

where a, b and k are constants.

Express p in terms of v, and find the work done by the gas in expanding from volume v_1 to v_2.

D.T.C.

13 (a) Evaluate $\int_0^1 (1+2x)^3\,dx$ and $\int_0^\pi \cos\left(2t-\frac{\pi}{3}\right)dt$

(b) The work W done by an expanding gas is given by $\dfrac{dW}{dV} = p$, where p is the pressure and V the volume. If for a certain gas $p(2V+1) = 100$, calculate the work done when this gas expands from $V = 27$ to $V = 38$.

14 A right circular cone has a height h and a mass M, and the radius of the base is r. The mass per unit volume is m. Show that its moment of inertia about the central axis of the cone is $\frac{3}{10}Mr^2$. (You may assume that the radius of gyration of a disc of radius a about an axis through its centre perpendicular to the disc is $\dfrac{a}{\sqrt{2}}$.)

D.T.C.

15 (a) Show that the second moment of area of a triangle of area A and perpendicular height h about the base is $\dfrac{Ah^2}{6}$ and, using the parallel axis theorem, deduce that the second moment of area about a parallel axis through the centroid is $\dfrac{Ah^2}{18}$.

(b) A lock gate has the shape of a symmetrical trapezium of width 30 m at the top and 20 m at the base, 10 m below. Find the second moment of the area about the top.

16 (a) When a gas expands from a volume V_1 to a volume V_2 the work done is given by $\int_{V_1}^{V_2} P \, dV$ where P is the pressure. The pressure and volume are connected by the law $PV^n = K$. If $P = 60$ when $V = 4$ and $P = 3.84$ when $V = 25$, find n correct to one decimal place, K correct to three significant figures and the work done in expanding from $V = 1$ to $V = 16$.

(b) Show that the function $x^3 + bx^2 + 3x$ has no turning points if b^2 is less than 9.

17 (a) The force F acting on a body at a distance s from a fixed point is given by $F = 2s + \dfrac{1}{s^2}$. Find the work done when the body moves from the position where $s = 0.5$ to that where $s = 2.5$. What is the mean value of the force during this movement?

(Work done is given by $\int F \, ds$)

(b) Use Simpson's rule with six strips to evaluate $\displaystyle\int_0^{1.5} \frac{1}{1+x^2} \, dx$.

18 Deduce, from first principles, an expression for the second moment of area of a rectangle about an edge in terms of its breadth and depth.

A symmetrical T-section is made up of two rectangles 5 cm × 1 cm. Determine the second moment of area about an axis perpendicular to the plane of the T and passing through the centroid.

19 Use Guldin's theorem to prove that the centroid of a semicircular area is at a distance $\dfrac{4r}{3\pi}$ from the centre, where r is the radius.

A rectangle ABCD has a semicircle described on the side BC.

If AB = CD = 10 cm and AD = BC = 6 cm, find the volume obtained for one revolution of this area about AD.

20 (a) With respect to second moments of area deduce:

  i the parallel axis theorem;
  ii the perpendicular axis theorem.

(b) Find the second moment of area, about an axis passing through the centroid perpendicular to the plane of the lamina of the H-section shown in fig. 74.

213   Miscellaneous exercises 9

Figure
74

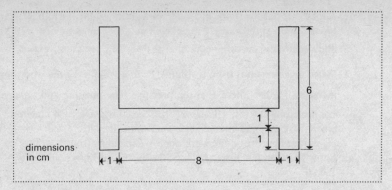

dimensions in cm

21 (a) State both Pappus' theorems concerning areas and volumes of revolution.
(b) Find the total surface area of the ring shown in half-section in fig. 75.

Figure
75

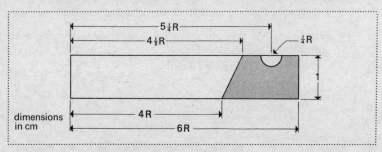

dimensions in cm

22 (a) ABCD is a rectangular area. If AB $= a$ and BC $= b$ show, from first
principles, that the second moment of area about CD is $\frac{1}{3}ab^3$.
(b) Find the second moment of area of the section in fig. 76 about the axis XY
and also about an axis through the centroid of the figure and parallel to XY.

Figure
76

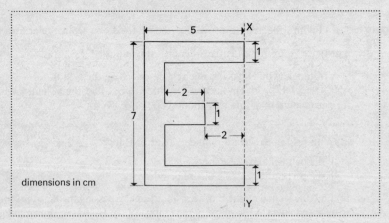

dimensions in cm

214 Applications of integration

23 (a) Establish the parallel axis theorem for second moments of area and state the perpendicular axis theorem.

  (b) Show that the second moment of area of a circle of radius r about an axis perpendicular to its plane through its centre is $\dfrac{\pi r^4}{2}$ and deduce the second moment of area about a diameter.

  (c) For the area between the two circles shown in fig. 77 find the second moment of area about the tangent XX, the diameter YY and also about the axis ZZ perpendicular to the figure through O.

Figure 77

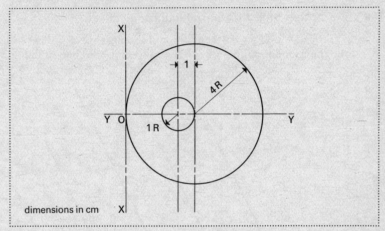

dimensions in cm

24 (a) Prove that the second moment of the area of a rectangle whose sides are of length l and b, about an axis through its centroid parallel to the sides of length b, is $\dfrac{bl^3}{12}$.

Figure 78

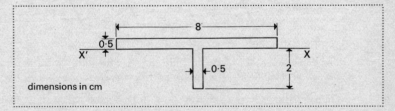

dimensions in cm

  (b) A T-section consists, as shown in fig. 78, of two rectangles; one 8 cm by 0·5 cm; the other 2 cm by 0·5 cm.

   Prove that the centroid of the section lies on the axis X'X.

   Calculate the second moment of area of the section about i the axis X'X, ii the axis of symmetry, iii an axis through the centroid perpendicular to the plane of the section.

10 Differential equations

10a Formation of a differential equation

The equation $y = 2x + k$ is the equation of a straight line of gradient 2. Plotting the graph for different values of k gives a set of parallel lines as in fig. 79.

Figure 79

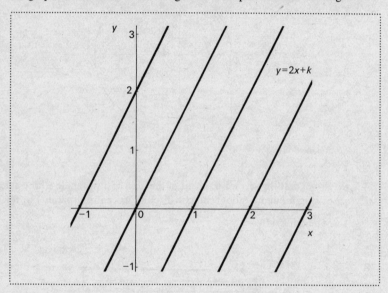

These lines are said to form a *family* and the equation $y = 2x + k$ is called the *equation of the family*. k is known as the *parameter* defining the curve.

Differentiating $y = 2x + k$ gives $\dfrac{dy}{dx} = 2$. This result which states that the gradient is 2 holds for every member of the family.

$\dfrac{dy}{dx} = 2$ is called a differential equation since it contains a derivative.

In the same way the equation $y = x^2 + k$ is the equation of a family of parabolas as shown in fig. 80.

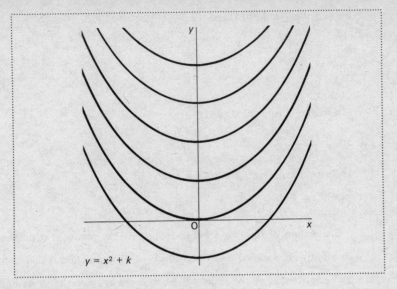

$$y = x^2 + k$$

Differentiating this equation with respect to x gives

$$\frac{dy}{dx} = 2x$$

This is the differential equation of the family of curves $y = x^2 + k$. It expresses the property that at every point on every curve of the family the gradient is twice the abscissa.

Example 1 Find the differential equation of the family of curves $y = ax^3$.

If $y = ax^3, \dfrac{dy}{dx} = 3ax^2$ and eliminating a from these two equations gives

$$a = \frac{y}{x^3} = \frac{1}{3x^2}\frac{dy}{dx}$$

Hence $x\dfrac{dy}{dx} = 3y$

Example 2 Find the differential equation of the family of curves $y = ax^2 + bx$.

We eliminate the arbitrary constants a and b.

$$y = ax^2 + bx \tag{10.1}$$

$$\frac{dy}{dx} = 2ax + b \tag{10.2}$$

$$\frac{d^2y}{dx^2} = 2a \tag{10.3}$$

Eliminating b from (10.1) and (10.2)

$$y = ax^2 + x\left(\frac{dy}{dx} - 2ax\right)$$

$$= x\frac{dy}{dx} - ax^2 \tag{10.4}$$

Eliminating a from (10.3) and (10.4)

$$y = x\frac{dy}{dx} - \frac{x^2}{2}\frac{d^2y}{dx^2}$$

This gives the differential equation

$$x^2\frac{d^2y}{dx^2} - 2x\frac{dy}{dx} + 2y = 0$$

Elimination of the two arbitrary constants has produced a differential equation with a second order derivative $\frac{d^2y}{dx^2}$. The equation is said to be of the *second order*, as the order of the equation is the order of its highest derivative. The *degree* of a differential equation is the degree of the highest derivative in the equation.

Thus $\left(\frac{d^2y}{dx^2}\right)^3 = 1 + \left(\frac{dy}{dx}\right)^4$ is of the second order and third degree.

In general the elimination of one arbitrary constant requires an equation of the first order and the elimination of two arbitrary constants requires an equation of the second order. The converse holds that the most general solution of a differential equation of the first order contains one arbitrary constant and the most general solution of an equation of the second order contains two arbitrary constants.

Example 3 The differential equation of a family of curves is given as $\frac{dy}{dx} = 3x$. Find the equation of the family.

The process in examples 1 and 2 is simply reversed.

$$\frac{dy}{dx} = 3x$$

Integration gives $\quad y = \frac{3x^2}{2} + C$

This is the equation of the family of curves. C is an arbitrary constant in the sense that $y = \frac{3x^2}{2} + C$ is a solution whatever the value of C.

Example 4 Find the equation of the family of curves whose gradient is given by $\dfrac{dy}{dx} = 3 - 2x$.

Find also the curve of the family which passes through the point (2, 3).

Since $\dfrac{dy}{dx} = 3 - 2x$

the family of curves is obviously $y = 3x - x^2 + C$ where C is the parameter.
   For the particular curve of the family passing through the point (2, 3) we get

$3 = 3 \times 2 - 2^2 + C$
$C = 1$

The particular curve is $y = 3x - x^2 + 1$.

Exercise 10a

In questions 1 to 8 find the differential equation of the family of curves.

1   $y = ax + 1$ 2 $4ay = x^2$ 3 $y = ax^4$

4   $y = ax^2 + b$ 5 $y = ax + \dfrac{1}{x}$ 6 $y = x^3 + ax$

7   $y = ax^3 + bx$ 8 $y = ax^3 + bx^2$

9   Find the equation of the family of curves whose gradients are given by the functions

   (a) $5x + 2$ (b) $2x + x^2$ (c) $2 \cos 2x$

   (d) $6x^2 - \dfrac{3}{x^2}$ (e) e^x (f) $\dfrac{x^2 + 1}{x}$

10   Find the equation of the curve whose gradient is given by $4x^3 - 2x$ and which passes through the point (1, 2).

11   Find the value of y when

   (a) $\dfrac{dy}{dx} = 3x^2 - 6x$ and $y = 0$ when $x = 1$.

   (b) $\dfrac{dy}{dx} = 3 - \dfrac{2}{x^3}$ and $y = 4$ when $x = 1$.

   (c) $\dfrac{dy}{dx} = e^{2x} - e^{-2x}$ and $y = 2$ when $x = 0$.

12　(a) If $\dfrac{ds}{dt} = 10-2t$ and $s = 40$ when $t = 2$, find an expression for s.

(b) If $\dfrac{dp}{dv} = 5v-2$ and $p = 6$ when $v = 1$, find an expression for p.

(c) If $\dfrac{ds}{dt} = 2+\dfrac{2}{5}t$ and $s = 12$ when $t = 5$, find s.

10b　Solution of differential equations

(a) The equation $\dfrac{dy}{dx} = f(x)$

An equation of this type can be solved immediately by integration.

Example 1　Solve the equation $\dfrac{dy}{dx} = 4e^{2x}+x^2$.

$$y = \int (4e^{2x}+x^2)\,dx+C$$

$$y = 2e^{2x}+\frac{x^3}{3}+C \quad \text{This is known as the general solution.}$$

Example 2　Solve the equation $x\dfrac{dy}{dx} = 2x+3$.

Here　$\dfrac{dy}{dx} = 2+\dfrac{3}{x}$

i.e.　$y = \int\left(2+\dfrac{3}{x}\right) dx+C$

$y = 2x+3\log_e x+C$ is the general solution.

(b) The equation $f(y)\dfrac{dy}{dx} = g(x)$

Here the variables x and y are separated,
　　Integrating both sides with respect to x gives

$$\int f(y)\frac{dy}{dx}\,dx = \int g(x)\,dx$$

or　$\int f(y)\,dy = \int g(x)\,dx$

　　Thus the left hand side is the integral of a function of y with respect to y and the right hand side is the integral of a function of x with respect to x.

The most direct way with an equation of this type is to write it

$$f(y)\,dy = g(x)\,dx$$

and integrate each side with respect to its own variable.

Example 1   Solve the equation $\dfrac{dy}{dx} = \dfrac{x^2 - 3x}{y+2}$.

We separate the variables and integrate

$$(y+2)\,dy = (x^2 - 3x)\,dx$$

$$\int (y+2)\,dy = \int (x^2 - 3x)\,dx$$

$$y^2 + 2y = \frac{x^3}{3} - \frac{3x^2}{2} + C \quad \text{where } C \text{ is an arbitrary constant.}$$

There is no point in writing constants on both sides of the equation as in
$$y^2 + 2y + A = \frac{x^3}{3} - \frac{3x^2}{2} + B \quad \text{since } B - A = C \text{ is equivalent to one constant}$$
only.

The arbitrary constant may be written in any form which happens to be convenient, such as $\log C$ or C^2 or e^C, to simplify the form of the solution.

Example 2   Solve the equation $\dfrac{dy}{dx} = 2x(y+3)$.

Separating the variables

$$\frac{dy}{y+3} = 2x\,dx$$

Integrating $\log_e (y+3) = x^2 + C$
$$y+3 = e^{x^2 + C} = e^C \times e^{x^2}$$

Replacing e^C by A, a new constant, the solution is

$$y+3 = Ae^{x^2}$$

Example 3   Solve the differential equation $\dfrac{dy}{dx} = y^2(4x - 1)$ given that $y = 1$ when $x = 2$.

Separating the variables $\dfrac{dy}{y^2} = (4x - 1)\,dx$

Integrating $-\dfrac{1}{y} = 2x^2 - x + C$

When $x = 2$, $y = 1$, $-1 = 8 - 2 + C$

$$C = -7$$

Thus $\dfrac{1}{y} = 7 + x - 2x^2$

This is a particular solution of the differential equation. Any solution that does not involve arbitrary constants (the constants of integration) is known as a *particular solution* or particular integral.

(c) The equation $\dfrac{d^2y}{dx^2} = f(x)$

This is a second order differential equation whose general solution can be obtained immediately by integrating twice in succession.

Example 1   Solve the equation $\dfrac{d^2y}{dx^2} = 12x^2 - 6x$.

Integrating $\dfrac{dy}{dx} = 4x^3 - 3x^2 + A$

Integrating again $y = x^4 - x^3 + Ax + B$

As the equation is of second order its solution contains two arbitrary constants A and B.

Example 2   A projectile is fired vertically upwards with an initial velocity u. Find the equation linking distance s above the point of projection with time t.

The acceleration is constant for the motion and is equal to g downwards. The differential equation of the motion is thus

$$\dfrac{d^2s}{dt^2} = -g$$

Integrating with respect to time

$$\dfrac{ds}{dt} = -gt + A$$

When $t = 0$, $\dfrac{ds}{dt} =$ initial velocity $= u$

$$u = A$$

Hence $\dfrac{ds}{dt} = -gt + u$

Integrating with respect to t again

$$s = -\frac{gt^2}{2} + ut + B$$

When $t = 0$ distance $s = 0$ and $B = 0$

Required equation is $s = ut - \frac{1}{2}gt^2$

Exercise 10b

Find the general solutions of the equations 1 to 20.

1   $\dfrac{dy}{dx} = e^x + 2x$ 2 $x\dfrac{dy}{dx} = 1 - x^2$

3   $\dfrac{dy}{dx} = \sin 2x + \cos 2x$ 4 $x^2\dfrac{dy}{dx} = 2 + x$

5   $x^3\dfrac{dy}{dx} = 2x + 3$ 6 $\dfrac{dy}{dx} = 1 - \cos x + \cos^2 x$

7   $xy\dfrac{dy}{dx} = 1 + x^2$ 8 $(x+2)\dfrac{dy}{dx} = y + 3$

9   $\dfrac{dy}{dx} + 2y - 3 = 0$ 10 $\dfrac{dy}{dx} = y(2x - 1)$

11   $x(y+1) + y(x+1)\dfrac{dy}{dx} = 0$ 12 $\dfrac{dy}{dx} = y$

13   $\dfrac{dv}{dt} = g - kv$ 14 $v\dfrac{dv}{dx} = 2x - 4$

15   $\dfrac{dN}{dt} = kN$ 16 $\dfrac{dN}{dt} = -kN$

17   $\dfrac{d^2y}{dx^2} = 4e^{-x}$ 18 $\dfrac{d^2y}{dx^2} = 6x^2 - \sin 2x$

19   $t\dfrac{d^2s}{dt^2} = 1$ 20 $\dfrac{d^2x}{dt^2} = \cos t$

223   **Solution of differential equations**

In examples 21 to 24 find the particular solutions satisfying the given conditions.

21    $\dfrac{dy}{dx} + y = 0$ and $y = 2$ when $x = 0$

22    $x^2 \dfrac{dy}{dx} = 1 + x$ and $y = 0$ when $x = 1$

23    $t \dfrac{dx}{dt} = x + tx$ and $x = 1$ when $t = 1$

24    $\dfrac{dr}{d\theta} + r \sin \theta = 0$ and $r = 3$ when $\theta = \dfrac{\pi}{2}$

25   For a body at temperature θ above its surroundings the rate of fall of temperature is given by the differential equation $\dfrac{d\theta}{dt} = -k\theta$ where k is a constant. Find the solution of this equation given that $\theta = 60$ when $t = 0$. Find the value of k if after 100 s, $\theta = 54$.

26   The current i in a circuit is given by the differential equation $E - L\dfrac{di}{dt} = Ri$ where E is the constant applied voltage, L the inductance, and R the resistance. Solve the equation given that $i = 0$ when $t = 0$.

27   The angular velocity ω of a flywheel under a constant braking torque N is given by the differential equation

$I\dfrac{d\omega}{dt} + N = 0$ where I is its moment of inertia.

Find ω in terms of the time t given that $\omega = \omega_0$ when $t = 0$.
Calculate the time to bring to rest from a speed of 60π rad s^{-1} a flywheel of m.i. 100 kg m^2 under a braking torque of 40 N m.

28   The retardation of a body travelling in a resisting medium is proportional to the square of its velocity. Show that the equation of motion may be represented by the equation $\dfrac{dv}{dt} + kv^2 = 0$ and that the solution is

$t = \dfrac{1}{k} \left[\dfrac{1}{v} - \dfrac{1}{v_0} \right]$ where v_0 is the initial velocity.

Find k if its velocity is reduced from its initial value of 200 m s^{-1} to 100 m s^{-1} in 30 s.

224   Differential equations

10c The differential equation of the family of curves $y = a \sin (nx + \alpha)$

$y = a \sin (nx + \alpha)$

(a and α are arbitrary constants determining particular curves)

Differentiating with respect to x gives

$$\frac{dy}{dx} = na \cos (nx + \alpha)$$

$$\frac{d^2 y}{dx^2} = -n^2 a \sin (nx + \alpha)$$

$$= -n^2 y$$

Hence the differential equation of the family is

$$\frac{d^2 y}{dx^2} + n^2 y = 0$$

Conversely the solution of $\frac{d^2 y}{dx^2} + n^2 y = 0$ may be said to be $y = a \sin (nx + \alpha)$.

Since it contains two arbitrary constants a and α it is the general solution. The solution can also be written in the form

$y = A \sin nx + B \cos nx$ with A and B as arbitrary constants.

Example 1   Solve the differential equations

i $\dfrac{d^2 y}{dx^2} + 25y = 0$ ii $\dfrac{d^2 s}{dt^2} + 4s = 0$

Comparing with $\dfrac{d^2 y}{dx^2} + n^2 y = 0$ we get the general solutions

i $y = a \sin (5x + \alpha)$ ii $s = a \sin (2t + \alpha)$

Example 2   Find the particular solutions of the equation $\dfrac{d^2 s}{dt^2} + 9s = 0$ which fit the conditions that

i when $t = 0$, $s = 0$ and $\dfrac{ds}{dt} = 12$

ii when $t = 0$, $s = 10$ and $\dfrac{ds}{dt} = 0$

The general solution is $s = a \sin (3t + \alpha)$
We write this in the form $s = A \sin 3t + B \cos 3t$

i When $t = 0$ $s = 0$

$0 = A \sin 0 + B \cos 0$ i.e. $B = 0$

$s = A \sin 3t$

Differentiating $\dfrac{ds}{dt} = 3A \cos 3t$

when $t = 0$, $\dfrac{ds}{dt} = 12 = 3A \cos 0$

$A = 4$

Here the particular solution is $s = 4 \sin 3t$.

ii $s = A \sin 3t + B \cos 3t$

When $t = 0$ $s = 10$

$10 = A \sin 0 + B \cos 0$ i.e. $B = 10$

Differentiating $\dfrac{ds}{dt} = 3A \cos 3t - 30 \sin 3t$

when $t = 0$, $\dfrac{ds}{dt} = 0 = 3A \cos 0 - 30 \sin 0$

$0 = 3A$ and $A = 0$

Hence the particular solution is $s = 10 \cos 3t$.

10d Simple harmonic motion

Suppose a particle P is moving along the straight line XOX' in a manner such that its acceleration is proportional to its distance x from O and is directed towards O. (See fig. 81.)

Figure 81

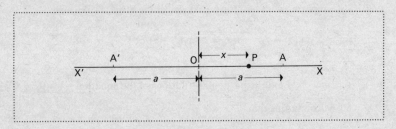

The velocity of the particle is given by $\dfrac{dx}{dt}$ and its acceleration by $\dfrac{d^2x}{dt^2}$.

The equation of motion of the particle is

$$\frac{d^2x}{dt^2} = -kx$$

As k is positive we may write this equation

$$\frac{d^2x}{dt^2} + n^2x = 0 \quad \text{where} \quad n^2 = k$$

The general solution of this equation is

$$x = a \sin(nt + \alpha) \quad \text{where } a \text{ and } \alpha \text{ are constants.}$$

The motion of P is known as *simple harmonic motion* (s.h.m.) and is oscillatory, moving between positions at a distance a on either side of the origin. The amplitude is a, $\dfrac{2\pi}{n}$ is the periodic time and $\dfrac{n}{2\pi}$ the frequency of the motion.

The acceleration is $-n^2x$ and has its greatest numerical value n^2a at A and A′ when $x = \pm a$.

The velocity $v = \dfrac{dx}{dt} = na \cos(nt + \alpha)$

$$= na\sqrt{[1 - \sin^2(nt + \alpha)]}$$

$$= na \sqrt{\left(1 - \frac{x^2}{a^2}\right)}$$

Hence $\qquad v^2 = n^2(a^2 - x^2)$

Thus the velocity v is a maximum ($= na$) when $x = 0$.

If the timing of the motion is started when P is at A (fig. 81) then when $t = 0$, $x = a$ and velocity $v = 0$, the solution then becomes $x = a \cos nt$.

Example A mass M is attached to a spring of stiffness S. If the system is set vibrating show that the motion is simple harmonic and find the periodic time of small oscillations.

If $M = 2$ kg and $S = 2$ N cm^{-1}, find the periodic time T. If the mass is displaced 40 cm beyond the equilibrium position and released, find its velocity when 10 cm from the equilibrium position.

Let x be the displacement of M from the equilibrium position.

Restoring force $= Sx$

By Newton's second law of motion

$$Sx = -M\frac{d^2x}{dt^2} \quad \text{where } \frac{d^2x}{dt^2} = \text{acceleration.}$$

Thus the equation of motion of the mass is

$$\frac{d^2x}{dt^2} + \frac{S}{M}x = 0$$

This is the equation of simple harmonic motion.

Comparing with $\dfrac{d^2x}{dt^2} + n^2 x = 0$ gives periodic time

$$T = 2\pi \sqrt{\dfrac{M}{S}}$$

$M = 2 \text{ kg}, S = 2 \text{ N cm}^{-1} = 200 \text{ N m}^{-1}$.

$$T = 2\pi \sqrt{\dfrac{2}{200}} = \dfrac{2\pi}{10} = 0\cdot628 \text{ s}$$

The velocity v is given by $v^2 = n^2(a^2 - x^2)$.

Here $\quad a = 0\cdot4 \text{ m}, x = 0\cdot1 \text{ m}$.

$$v^2 = \dfrac{200}{2}(0\cdot4^2 - 0\cdot1^2) = 15 \text{ m}^2 \text{ s}^{-2}$$

i.e. $\quad v = 3\cdot87 \text{ m s}^{-1}$

Exercise 10d

In questions 1 to 6 find the particular solutions of the following differential equations and give the periodic times of the oscillations.

1 $\dfrac{d^2y}{dx^2} + 4y = 0$ $\qquad$ when $x = 0, y = 0$ and $\dfrac{dy}{dx} = 10$

2 $\dfrac{d^2s}{dt^2} + s = 0$ $\qquad$ when $t = 0, s = 0$ and $\dfrac{ds}{dt} = 2$

3 $l\dfrac{d^2\theta}{dt^2} + g\theta = 0$ $\qquad$ when $t = 0, \theta = \theta_0$ and $\dfrac{d\theta}{dt} = 0$

4 $m\dfrac{d^2x}{dt^2} + sx = 0$ $\qquad$ when $t = 0, x = a$ and $\dfrac{dx}{dt} = 0$

5 $\dfrac{d^2\theta}{dt^2} + \dfrac{JCg}{Il}\theta = 0$ $\qquad$ when $t = 0, \theta = 0$ and $\dfrac{d\theta}{dt} = 4$

6 $4\dfrac{d^2y}{dx^2} + y = 0$ $\qquad$ when $x = 0, y = 4$ and $\dfrac{dy}{dx} = 2$

7 A particle moves in a straight line with s.h.m. about $x = 0$ as centre. Defining its position by $x = a \sin nt + b \cos nt$ find the constants a, b, and n if, when $t = 0$,

(a) $x = 0$ and $\dfrac{dx}{dt} = 12$. The periodic time is $\dfrac{\pi}{3}$.

(b) $x = 10$ and $\dfrac{dx}{dt} = 0$. The periodic time is 4π.

8 A spring carrying a mass M is displaced from its equilibrium position a distance x. If the restoring force is given by kx show that the motion is simple harmonic with equation

$$M\frac{d^2x}{dt^2} + kx = 0$$

If $M = 10$ g and $k = 1\cdot96$ N m^{-1}, find the periodic time of the motion and the frequency.

9 A mass M at the end of a cantilever is displaced a distance x from its equilibrium position. The restoring force due to the stiffness of the beam is given by

$$-\frac{3EI}{l^3}x$$

where E and I are constants for the beam and l is its length.

Show that the equation of motion on release is

$$\frac{d^2x}{dt^2} + \frac{3EI}{Ml^3}x = 0$$

Find the frequency of the oscillations.

10 At a certain port the depth between high and low water is 8 m. High water is at 02.00 h and low is at 08.15 h. If the motion of the tide is simple harmonic show that the height of the tide above the mean position in terms of the time t h from 02.00 h is given by $H = 4\cos\dfrac{4\pi}{25}t$.

Determine when the water level is 2 m below high water and find the rate in m min^{-1} at which the tide is then rising.

Miscellaneous exercises 10

1 (a) If $y = Ae^{px} + Be^{-px}$ show that $\dfrac{d^2y}{dx^2} = p^2y$.

(b) If $y = e^{mx}$ satisfies $\dfrac{d^2y}{dx^2} - \dfrac{3dy}{dx} + 2y = 0$ find the values of m.

U.L.C.I.

2   (a) If $y = (5-t) \cos \tfrac{1}{2}t$, prove that $4 \dfrac{d^2y}{dt^2} + y = 4 \sin \tfrac{1}{2}t$.

    (b) If $\dfrac{d^2y}{dx^2} = k(x-a)^2$, and both y and $\dfrac{dy}{dx}$ are zero when $x = 0$, show that
$12y = kx^2 (x^2 - 4ax + 6a^2)$.

E.M.E.U.

3   Find the differential equation to the family of curves $y^2 = 4ax$.

  Solve the differential equations

  (a) $x \dfrac{dy}{dx} = 2x^2 + 4$

  (b) $\dfrac{dN}{dt} = 10e^{-\frac{1}{2}t}$ and $N = 20$ when $t = 0$

4   (a) Evaluate i $\displaystyle\int_0^{\frac{1}{4}\pi} \sec^2\left(2x + \dfrac{\pi}{4}\right) dx$ ii $\displaystyle\int_2^3 \dfrac{dx}{x+2}$ iii $\displaystyle\int_1^2 x^2(2-x)^2 \, dx$

    (b) If $\dfrac{d^2x}{dt^2} = \sin 2t - 2 \cos 2t$ and $x = \tfrac{1}{2}, \dfrac{dx}{dt} = -\tfrac{1}{2}$ when $t = 0$, find x in terms
of t.

N.C.T.E.C.

5   (a) The equation of a curve is given by $y = A + Bx^2 + Cx^4$.
    Determine the values of A, B and C given that when

$$x = 0, \quad y = 1 \text{ and } \dfrac{d^2y}{dx^2} = -1$$

    and when $x = 1, \quad \dfrac{dy}{dx} = -\dfrac{2}{3}$

    (b) Solve the differential equation $\dfrac{d^2y}{dx^2} = e^x - e^{-x}$, given that when $x = 0$,

    $y = 3$ and $\dfrac{dy}{dx} = 1$.

U.L.C.I.

6   (a) Show that $y = Ae^x + Be^{2x}$ satisfies the differential equation

$$\dfrac{d^2y}{dx^2} - 3\dfrac{dy}{dx} + 2y = 0$$

    (b) Show that the differential equation of the family of curves $y^2 = 4ax + b$,
where a and b are constants, is

$$y \dfrac{d^2y}{dx^2} + \left(\dfrac{dy}{dx}\right)^2 = 0$$

7   (a) Differentiate with respect to x

   i $\log_e (2x^3)$ ii $\dfrac{1-\sin x}{1+\sin x}$ iii $e^{\frac{1}{2}x} \tan x$

   (b) Given that $\dfrac{d^2y}{dx^2} = -\dfrac{2}{x^2}$ find y in terms of x if $y = 0$ when $x = 1$, and

   $\dfrac{dy}{dx} = 3$ when $x = 1$.

8   (a) If $y = e^x \cos x$ show that $\dfrac{d^2y}{dx^2} - 2\dfrac{dy}{dx} + 2y = 0$.

   (b) The acceleration of a body at any instant t is given by $\dfrac{d^2s}{dt^2} = 9 + 6t$.

      When $t = 0$, $s = 12$ and when $t = 1$ the velocity is 6. Find s in terms of t.

   (c) A curve passes through the point $x = 3$, $y = 4$ and its gradient for all

      values of x is given by $\dfrac{dy}{dx} = 2x - \dfrac{x^2}{3}$. Find the equation of the curve.

9   (a) Differentiate with respect to x

   i $\log \dfrac{1}{x}$ ii $\sin\left(\dfrac{\pi}{6} - \dfrac{x}{2}\right)$ iii $\dfrac{x^2}{x^2+1}$

   (b) If $e^{x+y} - 2x = 0$ show that $\dfrac{dy}{dx} = \dfrac{1}{x} - 1$.

   (c) Show that $y = e^{3x} \cos 2x$ satisfies the equation

   $\dfrac{d^2y}{dx^2} - 3\dfrac{dy}{dx} + 4y + 6e^{3x} \sin 2x = 0$

                                                                    U.E.I.

10 (a) Obtain the series for $e^{-\frac{1}{2}x}$ in ascending powers of x as far as the term
      containing x^3.

      Use the series to obtain, correct to three decimal places, the values of

   i $e^{-0\cdot15}$ ii $\int_0^{0\cdot2} xe^{-\frac{1}{2}x}\,dx$

   (b) If $y = 10 \cos (nt + \alpha)$ in which $(nt + \alpha)$ is an angle measured in radians,

      $0 < \alpha < \dfrac{\pi}{2}$ and n is positive, verify that $\dfrac{d^2y}{dt^2} + n^2y = 0$.

      Obtain values for the constants n and α if $y = 6$ and $\dfrac{dy}{dt} = -16$ when
      $t = 0$.   *E.M.E.U.*

231   Miscellaneous exercises 10

11 (a) Find the differential equation of the family of curves $x^2 = 4ay + b$.
    (b) Solve the differential equations

$$\text{i} \frac{dy}{dx} = e^x - 2 \sin x \qquad\qquad \text{ii} \ (2x-1)\frac{dy}{dx} = y - 2$$

12 Solve the differential equations

$$\text{(a)} \ xy\frac{dy}{dx} = 1 + 2x - x^2 \qquad\qquad \text{(b)} \ \frac{d^2x}{dt^2} = 4e^{-t}$$

$$\text{(c)} \ x^2\frac{dy}{dx} = 2 - 4x \text{ and } y = 0 \text{ when } x = 1.$$

13 The differential equation of simple harmonic motion is $\dfrac{d^2x}{dt^2} + n^2x = 0$. Show
    that $x = A \sin nt + B \cos nt$ is a solution.

    Find the particular solution if $x = 0$ and $\dfrac{dx}{dt} = v_0$ when $t = 0$.

    A body moves with s.h.m. making 10 oscillations per second. The amplitude
    of the motion is 0·5 m. Find the maximum velocity and the maximum
    acceleration.

14 Show that $x = a \sin(\omega t + c)$ is a solution of the equation $\dfrac{d^2x}{dt^2} + \omega^2x = 0$.

    Find the solution of the differential equation $\dfrac{d^2x}{dt^2} + 9x = 0$ given that

    $x = 10$ and $\dfrac{dx}{dt} = 0$ when $t = 0$. What is the frequency of the oscillations?

15 (a) Find the differential equation of the family of curves represented by
        $x^2 + ky^2 = 1$.
    (b) Find the solutions of the differential equations

$$\text{i} \ \frac{dT}{d\theta} = \mu T \quad \text{given that } T = T_0 \text{ when } \theta = 0.$$

$$\text{ii} \ \frac{dy}{dx} = x(y+2)$$

16 (a) The equation of the family of concentric circles with centres at the origin
        is $x^2 + y^2 = a^2$.
        Show that the family may be represented by the differential equation
        $\dfrac{y}{x}\dfrac{dy}{dx} + 1 = 0$.

(b) Find the general solutions of the following differential equations

i $\dfrac{dy}{dx} = \dfrac{y^2+1}{2y}$ ii $EI\,\dfrac{d^2y}{dx^2} = w(l-x)$ where E, I, w, l are constants.

17   Solve the differential equations

(a) $\cos x\,\dfrac{dy}{dx} = 4$

(b) $L\,\dfrac{di}{dt}+Ri = E$ given $i = 0$ when $t = 0$.

(c) $v\,\dfrac{dv}{dx} = -g$ given $v = 10$ when $x = 0$.

18   Show that $s = A\cos \omega t + B\sin \omega t$ is a solution of the equation

$$\dfrac{d^2s}{dt^2}+\omega^2 s = 0$$

A body moves in a straight line with s.h.m. of amplitude 2 m and frequency 5 oscillations per second. Find the maximum velocity and maximum acceleration.

19   A body moves in a straight line with s.h.m. about $x = 0$ as centre. It moves through the centre with velocity 8 m s^{-1} and oscillates with amplitude 2 m. Express x in terms of the time t measured from the central position. Find the speed and acceleration of the body when one metre from the centre.

20   Solve the differential equations

(a) $\dfrac{dy}{dx} = 4x-e^{-x}$

(b) $\dfrac{d^2s}{dt^2} = 6t-4$ given $s = 10$ and $\dfrac{ds}{dt} = 4$ when $t = 0$.

(c) $\dfrac{dy}{dx} = (\cos x+e^{2x})\sec y$

11 Statistics

11a The mean and the standard deviation

Statistics deals with the collection and analysis of data. The data are normally sets of numerical values of a variable, the variable being the property or feature which is being measured.

One of the commonest ways of describing a set of numbers or values is to give the *average* or *mean* value. This is found by adding them all together and dividing by the number of values.

If the values of the variable are denoted by $x_1, x_2, x_3, \ldots, x_n$ then the mean $\bar{x}$ is defined by the equation

$$\bar{x} = \frac{x_1 + x_2 + \ldots + x_n}{n}$$

or $\quad \bar{x} = \dfrac{1}{n} \displaystyle\sum_{r=1}^{r=n} x_r$

where $\sum$ (sigma, the capital Greek letter S) means the sum of all such terms.

Thus, if a machine is set to cut 2-cm screws and a sample of eight are of lengths 2·02, 2·03, 1·96, 1·92, 1·90, 2·09, 2·03 and 1·92 cm, the mean length is

$$\bar{x} = \frac{2{\cdot}02 + 2{\cdot}03 + 1{\cdot}96 + 1{\cdot}92 + 1{\cdot}90 + 2{\cdot}09 + 2{\cdot}03 + 1{\cdot}92}{8}$$

$$= \frac{15{\cdot}87}{8} = 1{\cdot}98 \text{ cm to 2 D}$$

This average gives a standard to judge the machine's performance, but the average by itself is not sufficient to make valid comparisons say between two sets of data. A second machine may turn out screws varying between 1·50 cm and 2·50 cm in length and their mean may be approximately 2·0 cm but with such a variation in the lengths the machine is obviously not functioning properly. To estimate the spread or scatter the standard deviation is calculated.

Here the difference of each value from the mean is found and squared. The mean of these squared differences is found and its square root is called the *standard deviation* denoted by σ (sigma, the small Greek letter s).

It is defined by the equation

$$\sigma^2 = \frac{(x_1 - \bar{x})^2 + (x_2 - \bar{x})^2 + \ldots + (x_n - \bar{x})^2}{n}$$

$$= \frac{1}{n} \sum_{r=1}^{r=n} (x_r - \bar{x})^2$$

The square of the standard deviation, σ^2, is called the *variance*.

Example 1 Calculate the mean and the standard deviation of the following measurements in cm of the inner diameters of six cylinders.

2·25, 2·29, 2·36, 2·39, 2·31, 2·33.

$$\text{Mean} = \frac{2 \cdot 25 + 2 \cdot 29 + 2 \cdot 36 + 2 \cdot 39 + 2 \cdot 31 + 2 \cdot 33}{6} = \frac{13 \cdot 93}{6}$$

$$= 2 \cdot 32 \text{ cm}$$

$$\sigma^2 = \tfrac{1}{6} \big[(2 \cdot 25 - 2 \cdot 32)^2 + (2 \cdot 29 - 2 \cdot 32)^2 + (2 \cdot 36 - 2 \cdot 32)^2 + (2 \cdot 39 - 2 \cdot 32)^2$$
$$+ (2 \cdot 31 - 2 \cdot 32)^2 + (2 \cdot 33 - 2 \cdot 33)^2 \big]$$

$$= \tfrac{1}{6} \big[(-0 \cdot 07)^2 + (-0 \cdot 03)^2 + (0 \cdot 04)^2 + (0 \cdot 07)^2 + (-0 \cdot 01)^2 + (0 \cdot 01)^2 \big]$$

$$= \frac{0 \cdot 0125}{6} = 0 \cdot 002\,083$$

$$\sigma = \sqrt{0 \cdot 002\,083} = 0 \cdot 046 \text{ cm}$$

The arithmetic involved in the calculation of standard deviation can be tedious, especially if a calculating machine is not available. It is much easier to calculate the standard deviation about a trial mean and from it obtain the value about the true mean. The method depends upon the theorem which follows.

The second moment of a set of numbers about a value m is denoted by μ^2 and is defined as follows

$$\mu^2 = \frac{1}{n} \big[(x_1 - m)^2 + (x_2 - m)^2 + \ldots + (x_n - m)^2 \big]$$

$$= \frac{1}{n} \sum_{r=1}^{r=n} (x_r - m)^2$$

The variance is thus the second moment about the mean.

Corresponding to the parallel axis theorem for moments of inertia we have the result

$$\mu^2 = \sigma^2 + (\bar{x} - m)^2$$

This may be shown as follows:

$$\mu^2 = \frac{1}{n} \sum_{r=1}^{r=n} (x_r - m)^2$$

$$\mu^2 = \frac{1}{n} \sum_{r=1}^{r=n} x_r^2 - \frac{2m}{n} \sum_{r=1}^{r=n} x_r + \frac{m^2}{n} \sum_{r=1}^{r=n} 1$$

$$= \frac{1}{n} \sum_{r=1}^{r=n} x_r^2 - 2m\bar{x} + m^2$$

since $\dfrac{1}{n} \displaystyle\sum_{r=1}^{r=n} x_r = \bar{x}$ and $\displaystyle\sum_{r=1}^{r=n} 1 = n$

Also $\sigma^2 = \dfrac{1}{n} \displaystyle\sum_{r=1}^{r=n} (x_r - \bar{x})^2$

$$= \frac{1}{n} \sum_{r=1}^{r=n} x_r^2 - \frac{2\bar{x}}{n} \sum_{r=1}^{r=n} x_r + \frac{\bar{x}^2}{n} \sum_{r=1}^{r=n} 1$$

$$= \frac{1}{n} \sum x_r^2 - 2\bar{x}^2 + \bar{x}^2$$

$$= \frac{1}{n} \sum x_r^2 - \bar{x}^2$$

Thus $\mu^2 - \sigma^2 = \bar{x}^2 - 2m\bar{x} + m^2$

$$= (\bar{x} - m)^2$$

Example 2 The following are the masses in grams of twelve articles produced by the same machine. Find the mean and the standard deviation; 231, 234, 227, 235, 234, 230, 229, 227, 233, 226, 231, 228.

$m = 230$ is taken as a working mean and the differences from 230 and the second moment about 230 are found. The results are tabulated thus:

x Weight in grams	$x - 230 = d$	d^2
231	1	1
234	4	16
227	−3	9
235	5	25
234	4	16
230	0	0
229	−1	1
227	−3	9
233	3	9
226	−4	16
231	1	1
228	−2	4
Totals	$18 - 13 = 5$	107

Mean deviation from $230 = \dfrac{5}{12} = +0.42.$

$$\bar{x} = 230+0.42 = 230.42$$

$$\mu^2 = \frac{1}{12}\sum d^2 = \frac{107}{12} = 8.9167$$

$$= \sigma^2 + (\bar{x}-m)^2$$
$$= \sigma^2 + (0.42)^2$$

i.e. $\sigma^2 = 8.9167-0.1764 = 8.7403$

and $\sigma = \sqrt{8.7403} = 2.956$

Mean $= 230.42$ g

Standard deviation $= 2.96$ g

Exercise 11a

1   Find the mean and the standard deviation of the numbers 4, 5, 6, 7, 8, 9.

2   The breaking loads in kilonewtons of ten steel specimens were as follows: 30·2, 31·3, 32·2, 30·7, 31·9, 31·1, 30·8, 32·3, 31·6, 32·1. Find the mean breaking load and the standard deviation.

3   Find the second moment about the number 8 of the natural numbers 5 to 11. What is the mean and the standard deviation of these numbers?

4   The mean of a number of measurements is 16. The second moment about 12 is 30. Find the standard deviation and find the second moment about 18.

5   Eight shots are fired from a fixed gun and the ranges in metres are as follows: 3242, 3302, 3285, 3320, 3310, 3259, 3292, 3315. Find the mean range and the standard deviation, both to the nearest metre.

6   Of eight bags of potatoes, three contain 40 kg each, two contain 42 kg each, two more contain 46 kg each and the remaining bag contains 48 kg. Find the mean weight and the standard deviation.

7   Six samples from each of two machines turning out screws had lengths in centimetres as follows

Machine A	1·99	2·05	1·96	1·97	2·01	1·97
Machine B	2·04	2·06	1·95	1·95	2·05	1·97

Compare their means and standard deviations.

8   The lengths in centimetres of twelve rods are as follows. Find the mean value and the standard deviation.

5·23, 5·31, 5·33, 5·22, 5·29, 5·34, 5·35, 5·31, 5·27, 5·28, 5·32, 5·25.

11b Frequency distribution

When analysing large numbers of values of the variable it is normally impracticable to calculate mean and standard deviation by the previous methods where each value is dealt with separately. The range of values from smallest to largest is divided into a number of equal intervals and the numbers falling in any interval is known as the frequency. A table showing frequencies in each interval is known as a *frequency distribution*.

Example 1 The breaking strengths of fifty steel specimens are given below to the nearest 0·2 of a unit. Form a frequency distribution of the variable (i.e. the breaking strength).

32·4	32·6	32·0	32·8	33·4	33·6	33·0	33·2	32·2	32·8
32·6	32·8	33·0	32·6	33·0	33·2	32·6	32·8	33·0	32·4
32·8	33·0	33·0	33·2	33·4	32·8	33·0	32·4	32·8	33·0
32·2	33·6	32·2	33·4	33·2	32·8	32·4	32·8	33·0	32·8
33·0	33·4	32·6	33·2	32·6	32·8	32·8	33·2	33·0	33·0

Noting that the smallest value is 32·0 and the largest 33·6, a table is constructed with nine equal intervals of 0·2 and the following frequency distribution obtained.

Centre of interval	32·0	32·2	32·4	32·6	32·8	33·0	33·2	33·4	33·6
No. in interval, i.e. frequency	1	3	4	6	12	12	6	4	2

Mean and standard deviation

The mean and standard deviation are calculated by assuming that all the individual specimens in each interval have a value exactly equal to that of the centre of the interval.

The formulae for mean and standard deviation are modified as follows:

Let $x_1, x_2, x_3, \ldots, x_n$ be the values at the centres of the intervals and let $f_1, f_2, f_3, \ldots, f_n$ be the corresponding frequencies.

The total frequency = total number of values $N = f_1 + f_2 + f_3 + \ldots + f_n$.

$$\text{Mean } \bar{x} = \frac{1}{N} (f_1 x_1 + f_2 x_2 + f_3 x_3 + \ldots + f_n x_n)$$

$$= \frac{1}{N} \sum_{r=1}^{r=n} f_r x_r$$

Variance $\sigma^2 = \dfrac{1}{N}\left[f_1(x_1-\bar{x})^2 + f_2(x_2-\bar{x})^2 + \ldots + f_n(x_n-\bar{x})^2\right]$

$$= \frac{1}{N}\sum_{r=1}^{r=n} f_r(x_r-\bar{x})^2$$

The second moment of the distribution about m is given by

$$\mu^2 = \frac{1}{N}\sum_{r=1}^{r=n} f_r(x_r-m)^2$$

Also $\mu^2 = \sigma^2 + (\bar{x}-m)^2$ as before.

Example 2 Calculate the mean and the standard deviation of the breaking strengths of the fifty steel specimens in example 1.

A working mean $m = 32\cdot8$ is chosen and the calculations set out in table form.

Centre of interval x	Frequency f	$x-32\cdot8$ $=d$	fd	fd^2
32·0	1	−0·8	−0·8	0·64
32·2	3	−0·6	−1·8	1·08
32·4	4	−0·4	−1·6	0·64
32·6	6	−0·2	−1·2	0·24
32·8	12	0	0	0
33·0	12	0·2	2·4	0·48
33·2	6	0·4	2·4	0·96
33·4	4	0·6	2·4	1·44
33·6	2	0·8	1·6	1·28
Total	50		8·8 − 5·4 = 3·4	6·76

$$\bar{x} = m + \frac{\sum fd}{N}$$

$$= 32\cdot8 + \frac{3\cdot4}{50}$$

i.e. Mean $= 32\cdot8 + 0\cdot068 = 32\cdot87$

$$\mu^2 = \frac{\sum fd^2}{N} = \frac{6\cdot76}{50}$$

$$= 0\cdot1352 = \sigma^2 + (0\cdot068)^2$$

$$\sigma^2 = 0\cdot1352 - 0\cdot0046 = 0\cdot1306$$

Standard deviation $\sigma = \sqrt{0\cdot1306} = 0\cdot361$

Sheppard's correction

The assumption that each value is exactly that of the centre of the interval, when the values are distributed over the interval, has little effect on the value of the mean but it causes the calculated value of the standard deviation to be high.

An allowance may be made by subtracting $\dfrac{h^2}{12}$ from σ^2 where h is the size of the interval of the distribution.

This is known as Sheppard's correction.

Thus in example 2 above, the class interval $h = 0.2$.

$$\frac{h^2}{12} = \frac{0.04}{12} = 0.0033$$

The corrected standard deviation $= \sqrt{(0.1306 - 0.0033)} = \sqrt{0.1273}$
$$= 0.357$$

This correction only applies when the variable is continuous and the observed values are distributed over the interval. It must not be used when the variable takes only discrete values such as numbers of peas in pods, the integers, etc.

Coefficient of variation

This is used to compare the amount of scatter or dispersion of two distributions with different means and is given by $\dfrac{\sigma}{\bar{x}} \times 100\%$.

It is independent of the units used.

In example 2 above, using the corrected value for σ

$$\text{Coefficient of variation} = \frac{0.357 \times 100}{32.87} = 1.09$$

Exercise 11b

1 Find the means and standard deviations of the following distributions.

(a)

Centre of interval (kg)	27	28	29	Total
Frequency	2	7	1	10

(b)

Centre of interval (m)	3200	3300	3400	Total
Frequency	2	7	3	12

(c)

Centre of interval (cm)	275	276	277	Total
Frequency	5	12	3	20

2   The following table lists the javelin throws, measured to the nearest 5 m, recorded at a sports meeting. Find the mean and standard deviation.

Distance (m)	40	45	50	55	60	65	70	Total
No. of throws	2	4	10	12	7	4	1	40

3   In testing 25 ship's rockets the heights of the bursts were recorded as follows. Find the mean height and the standard deviation.

Height (m)	1215–1224	1225–1234	1235–1244	1245–1254	Total
No. of bursts	3	8	9	5	25

4   The length of 500 rods measured to the nearest centimetre are given in the following frequency table. Find the mean and the standard deviation. Find the corrected standard deviation using Sheppard's correction and calculate the coefficient of variation.

Centre of interval (cm)	61	62	63	64	65	66	67	68	69	70	71	72	73	Total
Frequency			4	9	14	26	48	76	88	78	58	48	32	16	3	500

5   In a series of tests the distance to take-off for a number of aircraft of the same type with equal loads was measured to the nearest five metres. Calculate the mean distance to take-off and the standard deviation from the following frequency table.

Distance to take-off (m)	240	245	250	255	260	265	270	275	280	Total
No. of aircraft		2	1	5	8	7	3	5	2	1	34

6   The lengths of a batch of rods are given in the following table, grouped in a frequency distribution with an interval of one centimetre.

Length (cm)	40	41	42	43	44	45	Total
Frequency	3	10	30	25	15	7	90

   Calculate the mean length and the standard deviation from the mean length, each to the nearest millimetre.

7   The breaking loads in kilonewtons to the nearest 0·2 kN of twenty specimens of a metal alloy are given in the following table.

Breaking load (kN)	25·0	25·2	25·4	25·6	25·8	26·0	Total
Frequency		2	2	5	7	3	1	20

   Calculate the mean and the standard deviation.

8   The breaking strengths of sixty steel specimens are given below, to the nearest 0·2 of a unit. Form a frequency distribution of the strengths and calculate the mean and the standard deviation of the distribution.

32·6	32·4	32·6	32·8	32·8	33·4	33·0	32·8
32·6	32·8	32·2	33·0	33·2	32·8	32·6	32·4
32·0	32·6	33·0	33·4	33·6	33·2	33·0	32·8
32·8	32·6	33·0	33·2	33·4	32·8	32·4	32·2
32·4	32·2	32·6	32·4	33·0	33·2	33·0	33·0
32·6	33·4	32·8	33·0	32·8	32·6	33·2	33·0
32·8	33·0	33·2	33·2	33·4	32·6	32·8	33·0
32·8	33·0	33·2	32·8				

9   The crushing loads in kilonewtons of forty wooden cubes are given in the following table. Calculate the mean and the standard deviation of this distribution.

Centre of interval	6·5	7·0	7·5	8·0	8·5	9·0	9·5	Total
Frequency	4	7	8	9	6	4	2	40

10   The following table gives the lengths of 200 metal rods as a frequency distribution grouped at 2 cm intervals. Find the mean length and the standard deviation from the mean using Sheppard's correction.

Centre of interval (cm)	60	62	64	66	68	70	72	74	76	78	Total
Frequency		4	8	23	35	62	44	18	4	1	1	200

11c   Graphical representation of numerical data

A frequency distribution can be represented graphically by plotting along the horizontal axis the intervals into which the range of values is split and erecting a rectangle on each interval with height equal to the corresponding frequency.

Figure 82

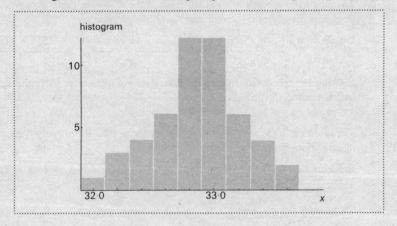

Such a diagram is called a *histogram*. Figure 82 shows the histogram for the frequency distribution in example 1 on page 238.

A frequency distribution may also be represented by a *frequency polygon* which is obtained by plotting the frequency to scale vertically about the centre of the corresponding interval on the horizontal axis. Figure 83 shows the frequency polygon for the same example.

Figure 83

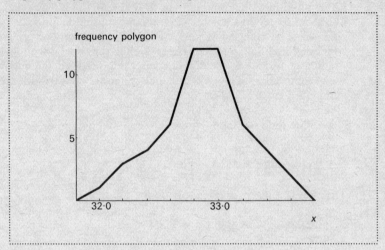

11d   Measures of position of a distribution

The histogram and frequency polygon show at a glance the position of the main group of values. Numerically the most important measure of position of a distribution is the mean. Two other measures of the central tendency of a distribution are the mode and the median.

The mode

This is the value of the variable which occurs most frequently, i.e. the value of x with the highest frequency. On a histogram it is the centre of the interval corresponding to the highest rectangle; on a frequency polygon it is the value corresponding to the highest point of the graph.

The median

This is defined as the middle number of the set when the numbers are arranged in order of magnitude. It is the value of x such that half the numbers are greater than x and half are less than x. If the number of values of the variable is even, the median may be taken as half way between the two central values.

Example The following frequency table gives the heights to the nearest centimetre of 100 schoolboys. Find i the mode and ii the median.

Height (cm)	152	153	154	155	156	157	158	159	160	161	162	163	164	165	Total
No. of schoolboys	1	2	3	4	6	8	16	23	14	11	6	4	1	1	100

i The value which occurs most frequently is 159 cm. This is the mode.

ii As there are an even number of observations the median is taken as the average of the 50th and 51st observations. There are forty observations in the range from 152 to 158 cm. Hence the 50th individual is the 10th in the next interval which is centred on 159 cm. The 23 heights in this interval are assumed to be spread evenly over the one-centimetre interval which stretches from 158·5 cm to 159·49 cm.

$$50\text{th observation} = 158\cdot5 + \frac{10}{23}\text{ cm}$$

$$51\text{st observation} = 158\cdot5 + \frac{11}{23}\text{ cm}$$

Thus the median may be taken to be

$$158\cdot5 + \frac{10\cdot5}{23} = 158\cdot5 + 0\cdot46 = 158\cdot96\text{ cm}$$

Cumulative frequency distribution

There is another method by which the median may be found and it is of special use in large distributions where the calculation of the middle value is tedious. A cumulative frequency table is constructed. The table gives, for each

Figure 84

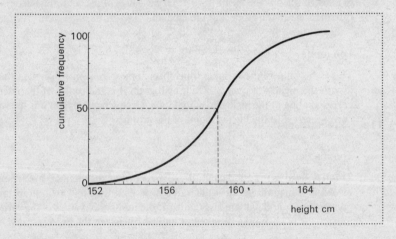

observation, the numbers of observations which are less than or equal to that value.

Thus for the heights of the 100 schoolboys in the example above the following table gives the cumulative distribution.

Centre of interval (cm)	Frequency f	Cumulative frequency
152	1	1
153	2	3
154	3	6
155	4	10
156	6	16
157	8	24
158	16	40
159	23	63
160	14	77
161	11	88
162	6	94
163	4	98
164	1	99
165	1	100

The cumulative frequency diagram is shown in fig. 84. Because of its typical shape it is sometimes called the *ogive curve*. The median is the value that cuts the observations in half. Thus for all practical purposes it can be taken as the value on the horizontal scale corresponding to $\dfrac{50+51}{2} = 50\cdot5$ on the vertical scale.

Note that when reading from the diagram the cumulative frequency numbers give totals up to the end of the interval.

In this case, and using a larger scale than in fig. 84, the median again comes to $158\cdot96$ cm.

11e   Measures of dispersion

To make valid comparisons between sets of data it is obviously not sufficient just to compare the means or medians. As pointed out in the beginning of the chapter the degree of spread or scatter of the distribution about a central value is important. This is normally measured by the standard deviation. Two other measures are: (a) the mean deviation from the mean (or it can be from the median in some cases), (b) the interquartile distance.

The mean deviation from the mean

The steps in the calculation are as follows
1 Calculate the mean, $\bar{x}$.
2 Calculate the difference between observed values and the mean, $(x_r - \bar{x})$.
3 Add all the results of 2 together, treating all values as positive.
4 Divide the total by N.

Thus mean deviation $= \dfrac{1}{N} \sum_{r=1}^{r=n} f_r |x_r - \bar{x}|$

Example 1 Find the mean deviation of the following distribution.

Centre of interval (cm)	23	25	27	29	31	33	Total
Frequency		1	4	10	12	6	2	35

Choose a working mean, 28 and denote the difference between this and the true mean by $\bar{e}$. The working can be set in a tabular form as follows.

x cm	f	$\begin{aligned}x-28\\=d\end{aligned}$	fd	$|x-(28+\bar{e})|$	$f|x-(28+\bar{e})|$
23	1	-5	-5	$5+\bar{e}$	$5+\bar{e}$
25	4	-3	-12	$3+\bar{e}$	$12+4\bar{e}$
27	10	-1	-10	$1+\bar{e}$	$10+10\bar{e}$
29	12	1	12	$1-\bar{e}$	$12-12\bar{e}$
31	6	3	18	$3-\bar{e}$	$18-6\bar{e}$
33	2	5	10	$5-\bar{e}$	$10-2\bar{e}$
Total	35		$\begin{aligned}40-27\\=13\end{aligned}$		$67-5\bar{e}$

$$\bar{e} = \frac{\sum fd}{N} = \frac{13}{35} = 0\cdot371$$

$$\text{Mean} = 28 + \bar{e} = 28\cdot37 \text{ cm}$$

$$\text{Mean deviation from the mean} = \frac{67 - 5 \times 0\cdot371}{35} = 1\cdot86 \text{ cm}$$

The interquartile distance

The median divides a distribution into halves with equal numbers of observations in each part. The quartiles divide each of these halves into two further equal parts, so that the two quartiles and the median divide the distribution into four equal parts with 25% of the observations in each. (See fig. 85.)

Fifty per cent of the observations fall between Q_2 (the upper quartile) and Q_1 (the lower quartile).

Figure
85

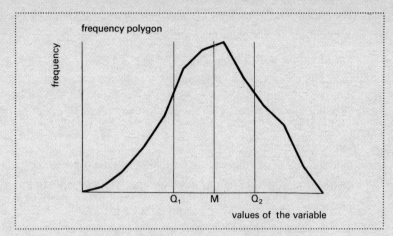

The distance $Q_2 - Q_1$ is known as the *interquartile distance*.

The semi-interquartile range $= \frac{1}{2}(Q_2 - Q_1)$ and this quantity is used as a measure of dispersion.

A distribution can be shown graphically as a cumulative frequency curve with the total frequency divided into 100 equal parts called *percentiles*. (See fig. 86.) The median is then the 50 percentile and the lower and upper quartiles the 25 and 75 percentiles.

Figure
86

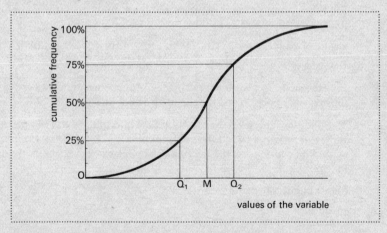

Example 2 Find the semi-interquartile range for the distribution in example 1 above.

The lower quartile is the value of the $\frac{1}{4}(35+1) = $ 9th individual
The median is the value of the $\qquad \frac{2}{4}(35+1) = $ 18th individual
The upper quartile is the value of the $\frac{3}{4}(35+1) = $ 27th individual

The 9th value is the 4th value in the interval 26 cm to 28 cm

$= 26 + \frac{4}{10} \times 2 = 26 \cdot 8$ cm $=$ lower quartile

The 27th value is the 12th in the interval 28 to 30 cm

$= 28 + \frac{12}{12} \times 2 = 30$ cm $=$ upper quartile

Semi-interquartile range $= \dfrac{30 - 26 \cdot 8}{2} = 1 \cdot 6$ cm

Exercise 11e

1   Find the median and the mode of the following frequency distribution.

Centre of interval	45	46	47	48	49	50	51	52	53	54	55	Total
Frequency		3	1	7	12	16	37	12	6	3	2	1	100

2   The breaking loads of eighty steel specimens are given in the table to the nearest 0·2 kN. Plot a histogram of the distribution. Prepare a cumulative frequency diagram and find the 25, 50, and 75 percentiles, showing them on the diagram. What is the semi-interquartile range?

Centre of interval (kN)	32·0	32·2	32·4	32·6	32·8	33·0	33·2	33·4	33·6	Total
Frequency	1	4	6	13	18	16	12	7	3	80

3   Find the upper and lower quartiles and the semi-interquartile range for the following distribution.

Centre of interval (cm)	95	97	99	101	103	105	Total
Frequency	2	12	26	24	8	3	75

4   Find the mode, median and mean of the distribution in example 3 above. Plot the frequency polygon and the cumulative frequency curve.

5   The following table gives the heights in 2 cm intervals of thirty-five plants of a variety of early peas. Find the mode, median and mean deviation from the mean. Plot a histogram and the cumulative frequency diagram for the distribution.

Centre of interval (cm)	23	25	27	29	31	33	Total
Frequency	1	4	10	12	6	2	35

6   Find the mode, median and mean of the following frequency distribution. Show these on the frequency diagram.

Centre of interval (cm)	31	32	33	34	35	36	37	38	39	40	41	42	43	44	Total
Frequency	52	81	63	54	30	22	14	10	6	3	1	2	1	2	341

7   The cumulative frequency diagram of a continuous distribution is represented by x^3 where x takes values from 0 to 1. Plot the cumulative frequency curve and show that the median is 0·794. Find the upper and lower quartiles and the semi-interquartile range.

8   Calculate the mean deviation from the mean of the following distribution.

Centre of interval	16	17	18	19	20	21	Total
Frequency	2	5	11	13	7	2	40

9   The following table gives the weights of 500 men in intervals of 20 units. Plot a cumulative frequency graph and find the median, the lower quartile and the upper quartile.

Weight	Under 100	100–119	120–139	140–159	160–179	180–199	200–219	220–239	Over 239	Total
No. of men	3	21	47	96	131	106	72	20	4	500

10 The following table shows the lengths of 400 screws measured to the nearest 0·001 cm. Plot a cumulative frequency graph and find the median, the upper and lower quartiles and the semi-interquartile range.

Length (cm)	0·993	0·994	0·995	0·996	0·997	0·998	0·999	1·000	1·001	1·002	1·003	Total
No. of screws	2	13	42	80	115	84	47	10	4	2	1	400

11f   Probability and sampling

With large volumes of data, the positions of distributions and their dispersions or spreads can be found by the measures outlined in the previous sections. A basic problem is the examination of results that can occur if only a sample of the possible observations can be taken. This requires some knowledge of probability.

Probability

Probability deals with the question as to the chance that a certain event will happen, and it is usually defined as follows:

If an event can happen in a ways and fail in b ways and all these events are equally likely to occur, then the probability of its happening is $\dfrac{a}{a+b}$ and the probability of its failing is $\dfrac{b}{a+b}$.

For example, if one ball is to be drawn from a bag containing three white balls and two red balls, the probability of drawing a white ball is $\frac{3}{5}$ and the probability of drawing a red ball is $\frac{2}{5}$.

Again, if a batch of 100 products has five defectives then the probability p of picking out a defective by one random selection is $p = \frac{5}{100} = 0.05$.

If in a number of trials an event has occurred n times and failed m times, we may estimate the probability of it occurring in the next trial as $\frac{n}{n+m}$. This is known as *empirical probability* as we are assuming that the only information is that of the past trials. Thus the probability of a man living through any one year, based upon past observations as recorded in a mortality table, is empirical probability.

On the probability scale a probability of 1 for an event to happen means that it is certain to happen, while a probability of 0 means that it will never happen. If p is the probability that an event will happen, then $1-p$ is the probability of its failing to happen.

Exclusive events

Events are said to be exclusive when the supposition that any one takes place is incompatible with the supposition that any other takes place.

In this case the probability that any one of them takes place is the sum of their probabilities.

Example 1 What is the probability of throwing a 2 or a 3 with an ordinary die (singular of dice)?

Probability of throwing a 2 is $\frac{1}{6}$
Probability of throwing a 3 is $\frac{1}{6}$
Probability of throwing a 2 or a 3 with one throw is $\frac{1}{6}+\frac{1}{6} = \frac{1}{3}$

Example 2 What is the probability of throwing not more than a 4 with one throw of a die?

Probability of throwing a 1 is $\frac{1}{6}$
Probability of throwing a 2 is $\frac{1}{6}$
Probability of throwing a 3 is $\frac{1}{6}$
Probability of throwing a 4 is $\frac{1}{6}$
Probability of throwing a 1, 2, 3 or 4 is $\frac{1}{6}+\frac{1}{6}+\frac{1}{6}+\frac{1}{6} = \frac{2}{3}$

Independent events

The probability that two independent events should both happen is the product of their separate probabilities of happening.

Example 3 What is the probability of throwing four heads in four tosses of a coin?

The probability of a head on first throw $= \frac{1}{2}$
The probability of a head on second throw $= \frac{1}{2}$
The probability of a head on third throw $= \frac{1}{2}$
The probability of a head on fourth throw $= \frac{1}{2}$

The probability of four heads $= \frac{1}{2} \times \frac{1}{2} \times \frac{1}{2} \times \frac{1}{2} = \frac{1}{16}$

Example 4   Ten per cent of a large batch of machine parts are defective, the defective parts being thoroughly mixed with the bulk. If two samples are drawn, what are the chances of picking i two defectives, ii no defectives, iii one defective only?

i   Probability of picking a defective on first pick $= \frac{1}{10}$
   Probability of picking a defective on second pick $= \frac{1}{10}$
   Probability of picking two defectives $\qquad = \frac{1}{10} \times \frac{1}{10} = 0.01$

ii   Probability of picking a good part on first pick $\quad = \frac{9}{10}$
   Probability of picking a good part on second pick $= \frac{9}{10}$
   Probability of picking no defectives $\qquad\qquad = \frac{9}{10} \times \frac{9}{10} = 0.81$

iii Probability of picking first a defective and then a good part
   $= \frac{1}{10} \times \frac{9}{10}$
   Probability of picking a good part then a defective
   $= \frac{9}{10} \times \frac{1}{10}$
   Probability of picking one defective only in two draws
   $= \frac{9}{100} + \frac{9}{100} = 0.18$

Note that all three probabilities $= 0.01 + 0.81 + 0.18 = 1.0$

The probability of an event happening r times exactly in n trials is $_nC_r\, p^r q^{n-r}$

Where p = probability of it happening.
$\qquad q = 1 - p$ = probability of it failing.

The probability of the event happening r times and failing $n-r$ times in any specified order

$= [p \times p \times p \ldots r \text{ times}] \times [q \times q \times q \ldots n-r \text{ times}]$
$= p^r q^{n-r}$

The number of different orders in which the r events could occur in n trials is $_nC_r$ and these are all equally probable.

Hence the probability of the event happening exactly r times in any order in n trials is $_nC_r\, p^r q^{n-r}$.

Expanding $(q+p)^n$ by the binomial theorem gives

$$(q+p)^n = q^n + {_nC_1}\, pq^{n-1} + {_nC_2}\, p^2 q^{n-2} + \ldots + {_nC_r}\, p^r q^{n-r} + \ldots + p^n$$

and the successive terms are the probabilities of the event happening exactly 0, 1, 2, 3 etc. times in n trials.

Example 5 Five pennies are thrown in an experiment. What are the probabilities of no heads occurring, one head only occurring, two heads occurring, etc.?

If this experiment is repeated sixty-four times, how many times could one expect groups of 0, 1, 2, 3, 4 and 5 heads to occur?

The probability of a coin falling and showing a head is $\frac{1}{2}$.

Thus $p = \frac{1}{2}, q = 1-\frac{1}{2} = \frac{1}{2}$ and $n = 5$

The probabilities are given by the successive terms of

$$\left(\frac{1}{2}+\frac{1}{2}\right)^5 = \left(\frac{1}{2}\right)^5 + 5\left(\frac{1}{2}\right)^4\left(\frac{1}{2}\right) + \frac{5\times4}{1\times2}\left(\frac{1}{2}\right)^3\left(\frac{1}{2}\right)^2 + \frac{5\times4\times3}{1\times2\times3}\left(\frac{1}{2}\right)^2\left(\frac{1}{2}\right)^3 +$$

$$+ \frac{5\times4\times3\times2}{1\times2\times3\times4}\left(\frac{1}{2}\right)\left(\frac{1}{2}\right)^4 + \left(\frac{1}{2}\right)^5$$

$$= \tfrac{1}{32}+\tfrac{5}{32}+\tfrac{10}{32}+\tfrac{10}{32}+\tfrac{5}{32}+\tfrac{1}{32}$$

i.e. the probabilities of 0, 1, 2, 3, 4 and 5 heads are $\frac{1}{32}, \frac{5}{32}, \frac{10}{32}, \frac{10}{32}, \frac{5}{32}, \frac{1}{32}$ respectively.

Multiplying these probabilities by 64 will give the expected numbers of times groups of 0, 1, 2 etc., heads would appear in each experiment. These are tabulated below and the probable numbers shown on the histogram of fig. 87.

Number of heads	0	1	2	3	4	5
Frequency	2	10	20	20	10	2

Figure 87

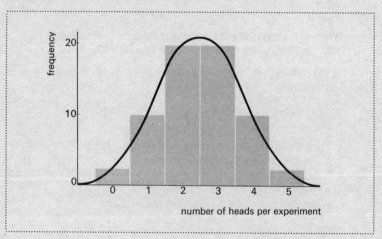

Example 6   Ten per cent of the articles from a certain machine are defective, the defective parts being thoroughly mixed with the bulk of parts. If three samples are taken what are the chances of picking

i   3 good products,   ii 1 defective,
iii 2 defective,   iv 3 defective?

What are the chances of at least one defective in a sample of 3?

Here $p = \frac{1}{10}, q = 1 - \frac{1}{10} = \frac{9}{10}, n = 3$

The successive terms of $\left(\frac{9}{10} + \frac{1}{10}\right)^3$ give the chances of 0, 1, 2 and 3 defectives respectively in a sample of 3.

i.e. $\left(\frac{9}{10}\right)^3 + 3\left(\frac{9}{10}\right)^2 \times \frac{1}{10} + 3\left(\frac{9}{10}\right)\left(\frac{1}{10}\right)^2 + \left(\frac{1}{10}\right)^3$

$= \frac{729}{1000} + \frac{243}{1000} + \frac{27}{1000} + \frac{1}{1000}$

i   Probability of 3 good products in the sample = 0·729
ii   Probability of 1 defective product in the sample = 0·243
iii   Probability of 2 defective products in the sample = 0·027
iv   Probability of 3 defective products in the sample = 0·001

The probability of at least one defective = 0·243 + 0·027 + 0·001 = 0·271

Exercise 11f

1   What is the chance that a card drawn at random from a normal pack of 52 cards is (a) an ace? (b) the ace of hearts?

2   A die is thrown twice. What are the probabilities that

(a) the first throw is a six?
(b) the first throw is not a six?
(c) the first throw is a six or a five?
(d) both throws are sixes?
(e) the total score is 8?
(f) the total score is 10?
(g) the total score is less than 13?

3   What is the chance of tossing a coin 5 times in a row and obtaining (a) 5 heads? (b) 4 heads or 4 tails? (c) 3 heads and 2 tails?

4   One of the numbers 1, 2, 3, 4, 5 is selected at random. What is the probability that the number will be (a) odd? (b) even?

5   A bag contains 5 red and 7 white balls. Find the probability of drawing 2 white balls in two draws

(a) the balls drawn not being replaced,
(b) the balls being replaced after each draw.

6   (a) In a box of 12 articles, 3 are damaged. If 3 are selected at random, what are the chances of picking 3 undamaged articles?
(b) In a large batch of articles 25% are damaged. If 3 are selected at random, what are the chances of picking 3 undamaged articles?

7   In a box of 10 springs, 3 are faulty. Two are chosen at random. What are the probabilities of picking (a) 2 faulty springs? (b) no faulty springs? (c) one faulty spring only?

8   In a large batch of articles 25% are damaged. If three articles are selected at random what are the chances that (a) no defectives are picked? (b) one defective is picked? (c) 2 defectives are picked? (d) 3 defectives are picked?

9   Four coins are thrown and the number of heads are counted. What are the chances of getting 0, 1, 2, 3 and 4 heads? If the experiment is repeated 128 times, draw up a table to show the numbers of throws one could expect to give 0, 1, 2, 3 and 4 heads.

10   In a box of seeds $\frac{1}{3}$ have pink flowers and $\frac{2}{3}$ have white flowers. In a row of 5 plants grown from these seeds, what are the probabilities of getting (a) all white flowers? (b) 2 pink flowers? (c) all pink flowers?
     In 243 rows of 5 plants approximately how many rows would you expect to contain (d) all pink flowers? (e) all white flowers?

11   Ten per cent of the articles from a machine are defective. In a sample of 12, what is the chance that there will be 3 defectives?

12   In a large batch of tyres 20% are substandard. In a random sample of 8 what are the chances that there will be (a) 2 substandard? (b) 4 substandard?

11g   Normal, binomial and Poisson distributions

Frequency curves

If, in a histogram, the class interval is decreased and the number of readings increased, the steps of the histogram would become smaller and smaller and its outline would approach more and more closely to a smooth curve. This limiting curve is known as a frequency curve. (See fig. 88.)
     If the distribution is symmetrical, the mean, median and mode all coincide. Otherwise the distribution is *asymmetrical* or *skew*. In this case the mean,

Figure
88

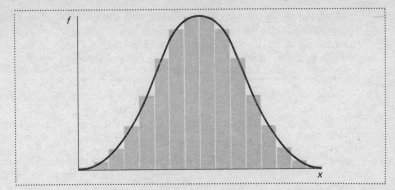

median and mode differ, with the median between the mean and the mode. (See fig. 89.)

For slightly asymmetrical distributions: mode−mean = 3(median−mean) approximately.

Figure
89

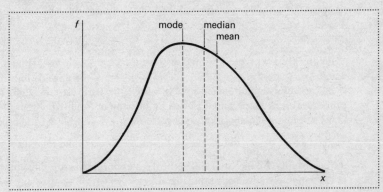

The skewness of a distribution is sometimes measured as

$$\frac{\text{mean}-\text{mode}}{\text{standard deviation}}$$

The normal distribution

This distribution gives the symmetrical bell-shaped curve which occurs with continuous natural distributions for large numbers of observations. The law of the distribution is represented by the equation

$$y = \frac{1}{\sigma\sqrt{(2\pi)}} e^{-(x-\bar{x})^2/2\sigma^2}$$

where $\bar{x}$ is the mean and σ is the standard deviation.

255 Normal, binomial and Poisson distributions

This is the law usually taken to represent the distribution of physical observations where the variation is due to small disturbing factors as likely to increase as to decrease the observations. The curve is known as the *normal distribution curve* and also as the Gaussian curve or the error curve.

Figure 90

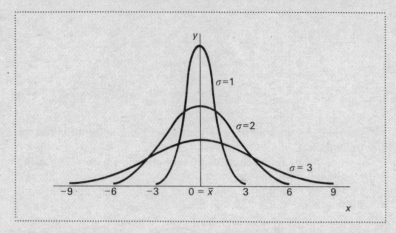

Figure 90 shows typical normal distribution curves with $\bar{x} = 0$ and $\sigma = 1$, 2 and 3.

Because of its importance the curve has been extensively tabulated and its properties investigated, and it is used for deductions for distributions that we know will follow closely the normal pattern.

The factor $\dfrac{1}{\sigma\sqrt{(2\pi)}}$ adjusts the vertical scale to make the total area under the curve unity. All the areas and vertical scale coordinates for different values of the variable x have been tabulated for $\sigma = 1$.

Figure 91

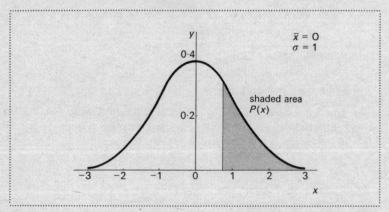

The normal probability table

This table gives values of the area shown shaded in fig. 91. Since the total area under the curve is unity, the area represented by $P(x)$ is expressed as a fraction of the total area under the curve. The curve is symmetrical.

Abscissa (x)	Area $P(x)$	Abscissa (x)	Area $P(x)$
0	0·5000	1·6	0·0548
0·1	0·4602	1·7	0·0446
0·2	0·4207	1·8	0·0359
0·3	0·3821	1·9	0·0287
0·4	0·3446	2·0	0·0228
0·5	0·3085	2·1	0·0179
0·6	0·2743	2·2	0·0139
0·7	0·2420	2·3	0·0107
0·8	0·2119	2·4	0·0082
0·9	0·1841	2·5	0·0062
1·0	0·1587	2·6	0·0047
1·1	0·1357	2·7	0·0035
1·2	0·1151	2·8	0·0026
1·3	0·0968	2·9	0·0019
1·4	0·0808	3·0	0·00135
1·5	0·0668	3·1	0·00097

The function $P(x)$ may be regarded as giving (a) the probability of obtaining in one trial a value greater than or equal to x or (b) the proportion of values greater than or equal to x in the population.

In general for a normal distribution $\bar{x}$ is not zero and σ is not unity but the table of values is still true for x in multiples of σ, measured from the mean, i.e. taking the abscissa in the table as $\dfrac{x-\bar{x}}{\sigma}$.

Notice that only 0·0027 of the distribution lies outside the limits of -3 and $+3$. This applies generally to any normal distribution. Only 0·27% of the values lie outside the range $\bar{x}-3\sigma$ to $\bar{x}+3\sigma$.

Thus the chance of an individual value lying outside the range $\bar{x}-3\sigma$ to $\bar{x}+3\sigma$ is about $\frac{1}{4}$%.

The limits $\bar{x}-3\sigma$ and $\bar{x}+3\sigma$ are often referred to as 99·75% confidence limits as with a normal distribution about 99·75% of the values will lie between these limits and in a random selection there is a 99·75% probability that the value chosen will lie between these limits.

Between $x = \bar{x}-2\sigma$ and $x = \bar{x}+2\sigma$ lie about 95·5% of the values with a normal distribution.

Between $x = \bar{x}-\sigma$ and $x = \bar{x}+\sigma$ lie about 68·2% of the values, while 50% lie between $\bar{x}+\frac{2}{3}\sigma$ and $\bar{x}-\frac{2}{3}\sigma$.

Example A certain size of torch battery has a mean life of 60 hours with a standard deviation of 2 hours. Of a batch of 200, how many will be expected to fail

i   between 56 and 62 hours?
ii after less than 58·4 hours?
iii After what length of time can one expect 180 of the 200 batteries to have failed?

i   Assuming normal distribution, 56 hours to 62 hours is from $\bar{x} - 2\sigma$ to $\bar{x} + \sigma$ (in fig. 91, from $x = -2$ to $x = 1$).
   The area outside these ranges from the normal probability table is $0·0228 + 0·1587 = 0·1815$.
   Thus the area between these ordinates is $1 - 0·1815 = 0·8185$.
   This is the fraction of the total number of batteries with an expected life between $\bar{x} - 2\sigma$ and $\bar{x} + \sigma$.

   Required number is $200 \times 0·8185 = 164$.

ii Less than 58·4 hours is less than $\bar{x} - 0·8\sigma$. The probability of values lying on or outside that value is $0·2119$.

   $200 \times 0·2119 = 42$, the expected number of failures.

iii 180 failures from $200 = 0·90$ of the total.
   From the table of values of $P(x)$ we have $0·0968 \simeq 0·1$ values of a normal distribution lie outside $\bar{x} + 1·3\sigma$, i.e. $60 + 1·3 \times 2$ hrs $= 62·6$ hours.
   Thus after 62·6 hours one can expect 180 of the batteries to have failed.

The binomial distribution

We have seen that the probabilities of an event happening exactly 0, 1, 2, ..., r, ... times in n trials are given by the successive terms of the binomial expansion of $(q + p)^n$ where p is the probability of it happening and $q = 1 - p$ is the probability of it not happening.

If we plot these probabilities vertically against the numbers of times the events happen and join the points by a curve we get frequency curves of the binomial distribution. (See fig. 87, p. 252.) For small values of n the curves are not symmetrical unless $p = q = \frac{1}{2}$, but as n is increased, the curves gradually approximate to a smooth symmetrical frequency curve of standard bell shape. It then becomes a very good approximation to the normal distribution curve.

The mean of the binomial distribution $\bar{x} = np$ and the standard deviation $\sigma = \sqrt{(npq)}$. It thus becomes a very easy distribution to work with.

Example A die is suspected of bias. In 720 throws 143 sixes were counted. Does this suggest bias?
   The probability of throwing a six with an unbiased die is $p = \frac{1}{6}$. It would be

very tedious to use the binomial expansion to obtain the probability of the numbers of sixes to be thrown. However, we know that the binomial distribution approximates to the normal distribution with n large.

$$\text{Mean } \bar{x} = np = 720 \times \tfrac{1}{6} = 120$$
$$\text{Standard deviation } \sigma = \sqrt{npq} = \sqrt{(720 \times \tfrac{1}{6} \times \tfrac{5}{6})} = 10$$
$$143 = \bar{x} + \tfrac{23}{10}\sigma = \bar{x} + 2{\cdot}3\sigma$$

From the normal probability table above, the chance of obtaining a value of $\bar{x} + 2{\cdot}3\sigma$ or over is about 0·011, i.e. only 1·1%.

Taking 5% as a significant level there is reasonable evidence that the die is biased.

The Poisson distribution

When the probability p of an event happening is very small, and the number of trials n is large then the formula for exactly r successes $_nC_r\, p^r q^{n-r}$ approaches the value of $\dfrac{e^{-np}(np)^r}{r!}$ or writing $np = m$, the probability of r successes is given by

$$P(r) = \frac{e^{-m}m^r}{r!}$$

The probabilities of 0, 1, 2, 3 etc. events occurring in n trials are given by

$$e^{-m}\left(1 + m + \frac{m^2}{2!} + \frac{m^3}{3!} + \ldots\right)$$

The mean value of the distribution $\bar{x} = m = np$
The standard deviation $\sigma = \sqrt{m} = \sqrt{(np)}$

These values are the same as for the binomial distribution if q is put equal to unity.

Example Two per cent of the articles produced by a machine are defective. Find the probabilities that in a sample of 40 taken at random there shall be 0, 1, 2 and 3 defectives. Assume a Poisson distribution.

$n = 40, p = 0{\cdot}02, m = 40 \times 0{\cdot}02 = 0{\cdot}8$

Probability of no defectives $\qquad e^{-m} = e^{-0{\cdot}8} = 0{\cdot}4493$

Probability of 1 defective $\quad e^{-m} \times m = 0{\cdot}4493 \times 0{\cdot}8 = 0{\cdot}3594$

Probability of 2 defectives $e^{-m} \times \dfrac{m^2}{2!} = 0{\cdot}3494 \times 0{\cdot}4 = 0{\cdot}1438$

Probability of 3 defectives $e^{-m} \times \dfrac{m^3}{3!} = 0{\cdot}1438 \times \dfrac{0{\cdot}8}{3} = 0{\cdot}0383$

11h   Standard error of the mean

If random samples of n individuals are taken from a large population, a series of means can be calculated. These will not be identical. The mean value of these sample means will be found to be very close to the true mean of the values as a whole. These sample means will be found to be more tightly packed round the true mean than are the individual values and the standard deviation of the means will be smaller than the standard deviation of the whole population.

If σ_n is the standard deviation of the means and σ is the standard deviation of the population it can be shown that

$$\sigma_n = \frac{\sigma}{\sqrt{n}}$$

This quantity $\dfrac{\sigma}{\sqrt{n}}$ is called the standard error of the mean.

Since individual means are unlikely to differ from the true mean by more than $\pm 3\sigma_n$, then the true mean is unlikely to differ from the sample mean by more than $3\sigma_n$, i.e. by more than $3\,\dfrac{\sigma}{\sqrt{n}}$. The larger the sample, the smaller is $\dfrac{\sigma}{\sqrt{n}}$ and the closer will be the value of the true mean. As it is impossible to measure σ it is usual to use the standard deviation of a large sample.

Example   If 400 samples of steel have a mean breaking strength of 474 MN m^{-2} with a standard deviation of 12·6 MN m^{-2} then the true mean is likely to lie between the following limits:

$$474 \pm 3 \times \frac{12 \cdot 6}{\sqrt{400}} = 474 \pm 1 \cdot 89$$

i.e. between 472 and 476 MN m^{-2}.

Exercise 11h

1   A normal distribution has a mean of 10 and a standard deviation of 2. Find the following areas under the normal distribution curve.

(a) From $x = 8$ to $x = 12$ (b) From $x = 8$ to $x = 13$
(c) From $x = 4$ to $x = 16$

2   The weights of the articles produced by a machine may be regarded as forming a normal distribution. What are the probabilities that the weight of one article selected at random will lie between (a) $\bar{x} - 1\frac{1}{2}\sigma$ and $\bar{x} + \sigma$, (b) $\bar{x} - 0 \cdot 8\sigma$ and $\bar{x} + 1 \cdot 2\sigma$?

3   Sixty tennis balls are dropped from a fixed height and the heights of bounce measured. The average height of bounce is 60 cm and the standard deviation 4 cm. How many of the tennis balls can be expected to bounce 66 cm or over?

4 One make of electric light bulb has an average life of 2000 hours and a standard deviation of 50 hours. Assuming a normal distribution what proportion of bulbs can be expected to burn for more than (a) 2050 hours, (b) 1925 hours?

5 If, from the large scale measurements of a species of insect, the length of the antennae is found to have a mean of 4·5 mm with a standard deviation of 1·2 mm, what is the probability that one of these insects picked up at random will have antennae between 3·9 mm and 5·7 mm? Assume a normal distribution.

6 Two per cent of the articles produced by a machine are defective. Find the probability that a random sample of 20 will contain no defectives assuming (a) a binomial distribution, (b) a Poisson distribution.

7 On average 3 % of the articles produced by a machine are defective. What are the probabilities of 0, 1, 2 and 3 defectives being in a sample of 40, using a Poisson distribution?

8 The average of 100 measurements of the lengths of screws produced by a machine is 2·008 cm and the standard deviation of the measurements is 0·38 cm. Between what limits does the mean almost certainly lie?

9 A certain make of electric light bulb has a mean life of 2000 hours with a standard deviation of 50 hours. What value would you expect for the standard deviation of the means of samples of 25 lamps.

10 Four hundred samples of a thread are found to have a mean breaking strength of 1·262 N with a standard deviation of 0·184 N. Between what limits will the true mean almost certainly lie, and between what limits will it lie with a probability of 0·95?

Miscellaneous exercises 11

1 The following numbers are the weights to the nearest gram of 80 parts produced by a machine and designed to weigh 120 g.

115	117	121	116	119	121	121	120	121	116
116	123	118	115	122	120	119	121	120	118
123	124	122	124	120	115	116	120	119	117
118	119	118	117	116	119	121	120	121	122
123	122	121	121	119	119	118	119	119	122
117	119	116	121	120	121	122	120	121	120
120	118	120	119	120	119	118	120	119	119
119	120	119	120	119	120	120	118	119	118

Form a frequency distribution of these values in one-gram intervals.
Find the mean and the mean deviation from the mean.
Draw a histogram showing the distribution.

2   Shells from an anti-aircraft gun are set so that they should all explode at the same height for a given angle of elevation of the gun.

In a test firing 50 rounds are fired at a fixed angle of elevation, with the following results:

Height of burst metres	5150–5250	5250–5350	5350–5450	5450–5550	5550–5650	5650–5750	5750–5850
Number of shells	2	3	7	14	12	8	4

Calculate the mean height of burst and the standard deviation.

D.T.C.

3   Two observers A and B each made ten measurements of the same quantity under the same conditions, and their readings were as follows:

A	323	322	325	323	324	323	321	324	323	322
B	325	324	321	324	323	326	325	323	324	325

Calculate the mean value and the standard deviation of each set of readings. State which observer you think is probably the more reliable, giving your reason.

D.T.C.

4   The marks out of ten obtained by 200 candidates in an examination are given in the following frequency table.

Mark	0	1	2	3	4	5	6	7	8	9	10	Total
Number of candidates	5	6	2	24	32	76	30	10	7	5	3	200

Find the mean, median and mode and show them on a frequency polygon. Plot a cumulative frequency curve and show on it the upper and lower quartiles.

5   (a) Define the mean, median and mode in relation to a given set of quantities.
(b) The percentage carbon content of 30 samples of a powder are given below.

1·2	2·4	2·8	3·5	3·2	3·2
3·9	3·0	2·6	3·1	2·2	2·8
2·9	3·8	2·3	3·0	2·7	3·2
2·4	3·4	3·5	1·6	4·0	4·4
4·0	4·7	3·5	3·4	4·2	1·8

i   Arrange these in a frequency distribution table with intervals from $0-1·4$, $1·5-$, $2·0-$, $2·5-$, etc.
ii Sketch the histogram.
iii Show the cumulative frequency both in the table and on the histogram.
iv Calculate the mean percentage carbon content.
v   Calculate the median percentage carbon content from your frequency distribution.

vi Calculate the approximate value of the mode.

vii Show the positions of the mean, median and mode on your sketch. Under what conditions of distribution are the mean, median and mode of the same value?

D.T.C.

6 A sample batch of 100 Hounsfield test specimens, of dimension 0.1785 ± 0.0003, for use in tensile experiments are turned off on an automatic lathe. The individual dimensions taken were distributed as shown in the table following.

Dimension	0·1782	0·1783	0·1784	0·1785	0·1786	0·1787	0·1788
Frequency	5	12	23	33	16	7	4

Calculate (a) the mean value, (b) the standard deviation for this sample.

D.T.C.

7 The crushing load in kilonewtons of 80 concrete cubes is given in the following frequency table, the values grouped to the nearest 0·5 kN.

Load (kN)	6	6·5	7	7·5	8	8·5	9	9·5	Total
Frequency	2	6	14	16	18	12	8	4	80

Prepare a cumulative frequency diagram and find the 25, 50 and 75 percentiles, showing them on the diagram.

What is the semi-interquartile range?

8 (a) Define the mean (M) and the standard deviation (σ) of n measurements x_1, $x_2, x_3, \ldots, x_n$ of a quantity.

If S is the root-mean-square deviation of these measurements from a number A (a 'fictitious mean'), prove that

$$\sigma^2 = S^2 - (M - A)^2$$

(b) Find the standard deviation of the following 20 measurements of the thickness, x mm, of steel sheet, taking 7·60 as a fictitious mean.

x	7·58	7·59	7·60	7·61	7·62	7·63
Frequency	1	2	4	6	5	2

D.T.C.

9 The number of hours of instruction required by a group of 200 men before they were successful in passing a practical test is given in the following frequency table.

30–40	40–50	50–60	60–70	70–80	80–90	hours
5	32	67	73	20	3	men

Draw a histogram of the frequency distribution and calculate the mean time to pass the test and the standard deviation.

D.T.C.

10 Five coins are thrown and the number of heads counted. Find the probabilities of getting 0, 1, 2, 3, 4 and 5 heads. If the experiment is repeated 256 times, draw up a table to show the numbers of times one could expect to obtain 0, 1, 2, 3 4 and 5 heads.
Show these results on a histogram.

11 (a) In a box of 10 springs, 3 are damaged. If two are chosen at random, what are the chances of picking i 2 faulty springs, ii no faulty springs?
(b) In a large batch of springs 30% are damaged. If two are chosen at random, what are the chances of picking i 2 faulty springs, ii no faulty springs?

12 Fatigue tests on a particular aircraft component were repeated 500 times. The failure times are given in the table below, to the nearest 100 hours.

Time (hours)	2300	2400	2500	2600	2700	2800
Parts failing	3	74	163	192	58	10

Calculate the mean time to failure and the standard deviation.

13 The average number of matches in the boxes of a certain make is 43 with a standard deviation of 2 matches. In a consignment of 60 000 boxes how many boxes will contain (a) less than 40 matches, (b) between 41 and 43 matches?

14 Three per cent of the articles produced by a machine are defective. Find the probabilities that in a sample of 50 taken at random there shall be 0, 1, 2 and 3 defectives. Use the Poisson distribution.

15 The breaking loads of metal specimens are given below to the nearest 0·2 kN. Form a frequency distribution of the loads and draw a histogram of the results. Calculate the mean and find the standard deviation corrected by Sheppard's rule. Find the coefficient of variation for the distribution.

20·6	20·4	19·8	20·8	20·2	20·4	20·4	20·6	20·0	20·4
20·4	20·8	20·4	20·6	20·4	20·2	20·6	20·4	20·2	20·8
20·2	20·4	20·6	20·4	21·0	20·4	20·6	20·2	20·4	20·0
20·6	20·0	20·2	20·4	20·6	20·6	20·4	20·8	20·2	20·6
20·0	20·4	20·6	20·8	20·0	20·4	19·8	20·4	20·6	20·4
20·2	20·6	20·0	20·2	20·6	20·4	20·6	20·2	20·0	20·6

16 (a) A normal distribution has a mean of 20 and a standard deviation of 3. Find the area under the distribution curve from $x = 17$ to $x = 24·5$.
(b) A make of light bulb has an average life of 1800 hours and a standard deviation of 60 hours. Assuming a normal distribution, what number of bulbs from a batch of 120 can be expected to fail after 1700 hours? How many should last more than 1860 hours?

12 Complex numbers

12a Definition of a complex number

In solving quadratic equations by the formula $\dfrac{-b \pm \sqrt{(b^2 - 4ac)}}{2a}$, the term $b^2 - 4ac$ is occasionally negative giving roots containing a term $\sqrt{(-k)}$ where k is positive. Since all squares, whether of positive or negative quantities, are positive it follows that $\sqrt{(-k)}$ cannot represent any positive or negative quantity. It is on this account called an *imaginary* quantity, although the choice of word is unfortunate.

Finding a meaning for imaginary quantities extends further our ideas on numbers.

Consider the vector $\overline{AB}$ in fig. 92. A vector is a line where regard is paid both to its magnitude and its direction.

Figure 92

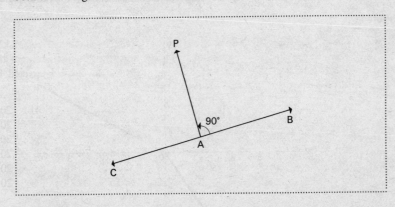

We know that a minus sign in front of a vector reverses it, that is it turns it through 180°.

Thus $-\overline{AB} = \overline{AC}$

We invent an operator j which when placed in front of a vector rotates that vector through 90° anticlockwise.

The symbol i is also used, but as i is also used to denote electric current, j is preferred in engineering.

Thus $\quad j\,\overline{AB} = \overline{AP}$

Operating again by j

$$j(j\,\overline{AB}) = j\,\overline{AP} = \overline{AC}$$

i.e. $\quad j^2\,\overline{AB} = \overline{AC}$

But $\quad \overline{AC} = -\overline{AB}$

i.e. $\quad j^2\,\overline{AB} = -\overline{AB}$

Thus $\quad j^2 = -$

We usually introduce 1 and say $j^2 = -1$.

No error is introduced by doing this since we could multiply through by 1 at any stage without affecting the result.

Again we can write $j = \sqrt{(-)}$ or $j = \sqrt{(-1)}$.

In this sense j is a square root of -1. The main thing to remember in operations involving j is that j^2 in multiplication must be replaced by -1.

Since $ja \times ja = j^2 a^2 = (-1)a^2 = -a^2$ it follows that $\sqrt{(-a^2)} = ja$ and it is only necessary to use one expression j with imaginary quantities.

A number in the form $a + jb$ is known as a *complex* number with a the *real part* and jb the *imaginary part*.

Complex numbers may be represented geometrically in the following way.

Figure
93

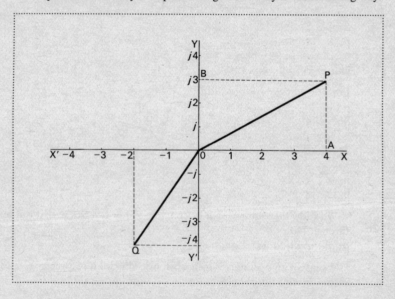

12b   The Argand diagram

Consider the two axes XOX′ and YOY′ at right angles as in fig. 93.

A real number is represented by a length along OX or a point on it. Positive values are on OX and negative values on OX′.

A purely imaginary number is measured along OY if positive and OY′ if negative.

Thus in fig. 93 point A or the vector $\overline{OA}$ represents the real number 4.

Point B or the vector $\overline{OB}$ represents the imaginary number $j3$ (or $3j$). The

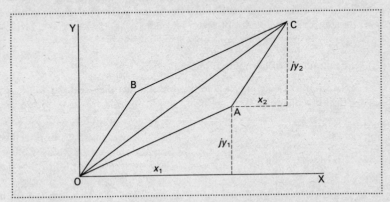

Figure
94

complex number $4+j3$ is represented by point P or by the sum of the vectors $\overline{OA}+\overline{OB}$, i.e. by their resultant $\overline{OP}$. Point Q or vector $\overline{OQ}$ represents the complex number $-2-j4$. XOX′ is called the real axis, YOY′ the imaginary axis. The diagram is known as an *Argand diagram* and P is called the *Argand point* of the complex number $4+j3$ (or $4+3j$).

If two complex numbers are equal their real parts are equal and their imaginary parts are equal.

For if $a+jb = c+jd$
$$a-c = j(d-b)$$
Squaring: $(a-c)^2 = j^2(d-b)^2 = -(d-b)^2$
$(a-c)^2+(d-b)^2 = 0$ i.e. the sum of the two squares is zero.
Thus each square is zero, i.e. $a = c$ and $b = d$.

We can see that this must be so from the Argand diagram since two equal numbers must be represented by the same point on the diagram and so must have their real parts equal and their imaginary parts equal.

12c   Addition, subtraction, multiplication and division

Addition of complex numbers

To add two complex numbers we add their real parts and add their imaginary parts, treating j as obeying the ordinary laws of algebra.

Thus $(x_1 + jy_1) + (x_2 + jy_2) = x_1 + x_2 + j(y_1 + y_2)$

The rule may be demonstrated on the Argand diagram.

Let $\overline{OA}$ represent the number $x_1 + jy_1$ and $\overline{OB}$ $x_2 + jy_2$. (See fig. 94.) From A draw AC equal and parallel to OB so that $\overline{AC}$ represents $x_2 + jy_2$.

Then $\overline{OA} + \overline{AC} = \overline{OC}$

From the diagram the abscissa of C is $x_1 + x_2$ and its ordinate is $y_1 + y_2$ so that $\overline{OC}$ represents the complex number $x_1 + x_2 + j(y_1 + y_2)$.

Thus complex numbers follow the ordinary law of algebraic addition and follow the ordinary law of vector addition.

Subtraction of complex numbers

To subtract two complex numbers we subtract their real parts and subtract their imaginary parts.

Figure
95

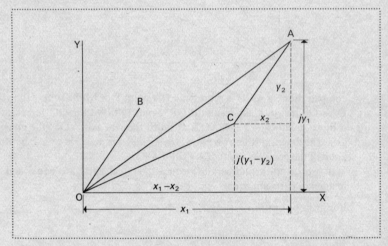

Let $\overline{OA}$ represent $x_1 + jy_1$ and $\overline{OB}$ $x_2 + jy_2$. (See fig. 95.) From A draw AC equal and parallel to OB but in the opposite direction so that $\overline{AC} = -\overline{OB}$.

Then $\overline{OA} - \overline{OB} = \overline{OA} + \overline{AC} = \overline{OC}$

From the diagram the abscissa of C is $x_1 - x_2$ and its ordinate is $y_1 - y_2$ so that $\overline{OC}$ represents the complex number $x_1 - x_2 + j(y_1 - y_2)$.

Thus we see that subtraction of complex numbers may be defined either by the ordinary laws of algebraic subtraction or by the laws for subtraction of vectors.

Example 1   Evaluate i $(4+j5)+(6+j)+(2-j3)$,
　　　　　　　ii $(2·5+j3·3)-(3-j4·7)$

i $(4+j5)+(6+j)+(2-j3) = 4+6+2+j(5+1-3)$
　　　　　　　　　　　　　$= 12+j3$

ii $(2·5+j3·3)-(3-j4·7) = 2·5-3+j(3·3--4·7)$
　　　　　　　　　　　　　$= -0·5+j8$

Multiplication of complex numbers

Complex numbers are multiplied together in the normal way, j^2 being replaced by -1 whenever it occurs.

Thus $j^2 = -1, j^3 = j^2 \times j = -j$,
　　　$j^4 = (j^2)^2 = (-1)^2 = 1, j^5 = j^4 \times j = j$, etc.

Example 2   Evaluate i $(a+jb)(c+jd)$, ii $(3+j4)(2-j)$

i $(a+jb)(c+jd) = ac+ajd+jbc+jbjd$
　　　　　　　　　$= ac+j(ad+bc)+j^2bd$
　　　　　　　　　$= ac-bd+j(ad+bc)$

ii $(3+j4)(2-j) = 6+j8-j3-j^24$
　　　　　　　　　$= 10+j5$

Conjugate complex numbers

Numbers of the form $a+jb$ and $a-jb$ are said to be conjugate. Their product is a real number.

$(a+jb)(a-jb) = a^2+jba-jab-j^2b^2$
　　　　　　　$= a^2+b^2$

Note that a conjugate is formed by changing the sign of the imaginary part only.

Division of complex numbers

In evaluating a quotient the denominator is made a real number by multiplying both denominator and numerator by the conjugate of the denominator.

For example $\dfrac{2+j3}{3-j4} = \dfrac{2+j3}{3-j4} \times \dfrac{3+j4}{3+j4}$

$$= \frac{6+j(9+8)+j^212}{3^2+4^2} = \frac{-6+j17}{25}$$

$$= -0·24+j0·68$$

Example 3 Simplify $\dfrac{3-j}{5+j2}+\dfrac{2+j3}{4-j3}$

$$\frac{3-j}{5+j2} = \frac{3-j}{5+j2}\times\frac{5-j2}{5-j2} = \frac{15-2-j(5+6)}{25+4}$$

$$= \frac{13}{29}-j\frac{11}{29} = 0{\cdot}448-j0{\cdot}379$$

$$\frac{2+j3}{4-j3} = \frac{2+j3}{4-j3}\times\frac{4+j3}{4+j3} = \frac{8-9+j(12+6)}{16+9}$$

$$= -\frac{1}{25}+j\frac{18}{25} = -0{\cdot}040+j0{\cdot}720$$

$$\frac{3-j}{5+j2}+\frac{2+j3}{4-j3} = 0{\cdot}448-0{\cdot}040+j(0{\cdot}720-0{\cdot}379)$$

$$= 0{\cdot}408+j0{\cdot}341$$

Exercise 12c

1 Indicate on a diagram vectors representing the following numbers:

(a) $3+j5$, (b) $4-j3$, (c) $-2+j3$, (d) $-4-j$

2 If z denotes $4+j3$ and $\bar{z}$ its conjugate, mark on a diagram the Argand points of the following numbers.

$z, \bar{z}, z+\bar{z}, z-\bar{z}, z+1+j2, z-1-j2$

Evaluate the following, putting in the form $x+jy$.

3 $(2+j3)(2+j)$ $\qquad$ 4 $(2+j)^2$ $\qquad$ 5 $\dfrac{3}{j}$

6 $\dfrac{3}{j^2}$ $\qquad$ 7 $\dfrac{3}{j^3}$ $\qquad$ 8 $(8-j4)(-7+j2)$

9 $(5+j13)(6+j2)$ $\qquad$ 10 $(2+j)^4+(2-j)^4$

11 $(1+j)^2+(1-j)^2$ $\qquad$ 12 $\dfrac{1+j}{1-j}$

13 $\dfrac{5+j3}{2+j3}$ $\qquad$ 14 $\dfrac{-7-j9}{5-j12}$ $\qquad$ 15 $\dfrac{-5+j8}{-10-j10}$

16 Find the real and the imaginary parts of:

(a) $\dfrac{1}{3+j2}$, (b) $\dfrac{2j}{(2-j)^2}$, (c) $\dfrac{106}{7-j2}$

17 Simplify to the form $x+jy$

(a) $(1+j2)^2(3-j5)$ (b) $\dfrac{1+j}{1-j2}+(1+j\sqrt{3})^3$ (c) $\left(\dfrac{-1+j\sqrt{3}}{2}\right)^4$

18 Solve (a) $(2-j3)x = 5-j3$, (b) $x^2+2x+6 = 0$

19 If Z is the Argand point of the complex number $4+j3$, what complex number is represented by

(a) the image of Z in OX, (b) the image of Z in OY?

20 The Argand points $2+j3$ and $4-j$ are opposite corners of a square. What complex numbers are represented by the other corners?

12d   The polar form of a complex number

Let (x, y) and (r, θ) be the cartesian and polar coordinates of P. (See fig. 96.) Denoting the complex number $x+jy$ by z,

$$z = x+jy$$
$$= r\cos\theta + jr\sin\theta$$
$$= r(\cos\theta + j\sin\theta)$$

$r(\cos\theta + j\sin\theta)$ is called the *polar form* of the complex number z and it may be written $r\underline{/\theta}$.

Figure 96

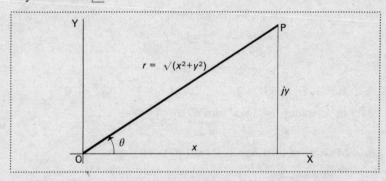

r is called the *modulus* of the complex number z and denoted by mod z or $|z|$.

Thus $|z| = r = \sqrt{(x^2+y^2)}$

θ is called the *argument* or *amplitude* of z and is written as arg z.

Thus $\arg z = \theta = \tan^{-1}\dfrac{y}{x}$

In finding θ care must be taken to select the correct one of two possible values for $\tan^{-1}\dfrac{y}{x}$. It is best to use a sketch to find the quadrant in which z lies. In stating the value of θ the least numerical value is chosen, i.e. the angle between $-\pi$ and π. This is known as the principal value.

Example Find the modulus and argument of the complex number $z = 6 - j3$ and express z in polar form.

Figure 97

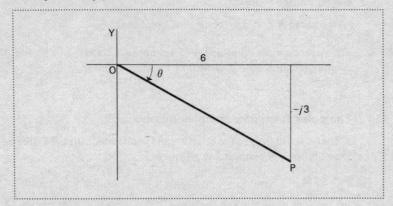

The complex number is represented by OP in fig. 97.

The modulus $|z| = \sqrt{(6^2 + 3^2)} = \sqrt{45} = 6{\cdot}71$

The angle θ is in the fourth quadrant.

$$\tan \theta = -\frac{3}{6} = -\frac{1}{2}$$

$$\theta = -26°\ 34'$$

Hence $|z| = 6{\cdot}71$, $\arg z = -26°\ 34'$

In polar form $6 - j3$ is written $6{\cdot}71\underline{/-26°\ 34'}$.

12e Multiplication and division in polar form

Multiplication

Let the complex numbers be

$$z_1 = r_1\underline{/\theta_1} = r_1(\cos\theta_1 + j\sin\theta_1)$$
$$\text{and} \quad z_2 = r_2\underline{/\theta_2} = r_2(\cos\theta_2 + j\sin\theta_2)$$
$$z_1 z_2 = r_1 r_2(\cos\theta_1 + j\sin\theta_1)(\cos\theta_2 + j\sin\theta_2)$$
$$= r_1 r_2[\cos\theta_1\cos\theta_2 + j\cos\theta_1\sin\theta_2 + j\sin\theta_1\cos\theta_2 + j^2\sin\theta_1\sin\theta_2]$$

$$z_1 z_2 = r_1 r_2 [(\cos \theta_1 \cos \theta_2 - \sin \theta_1 \sin \theta_2) + j(\sin \theta_1 \cos \theta_2 + \cos \theta_1 \sin \theta_2)]$$
$$= r_1 r_2 [\cos (\theta_1 + \theta_2) + j \sin (\theta_1 + \theta_2)]$$
$$= r_1 r_2 \underline{/\theta_1 + \theta_2}$$

Hence to multiply two complex numbers the rule is to multiply their moduli and add their arguments.

For example   i $3\underline{/30°} \times 2\underline{/40°} = 6\underline{/70°}$

ii $5\underline{/52°} \times 3\underline{/-43°} = 15\underline{/9°}$

Division

$$\frac{z_1}{z_2} = \frac{r_1(\cos \theta_1 + j \sin \theta_1)}{r_2(\cos \theta_2 + j \sin \theta_2)}$$

$$= \frac{r_1}{r_2} \frac{(\cos \theta_1 + j \sin \theta_1)}{(\cos \theta_2 + j \sin \theta_2)} \times \frac{(\cos \theta_2 - j \sin \theta_2)}{(\cos \theta_2 - j \sin \theta_2)}$$

$$= \frac{r_1}{r_2} \frac{(\cos \theta_1 \cos \theta_2 - j^2 \sin \theta_1 \sin \theta_2) + j(\sin \theta_1 \cos \theta_2 - \cos \theta_1 \sin \theta_2)}{\cos^2 \theta_2 - j^2 \sin^2 \theta_2}$$

$$= \frac{r_1}{r_2} [\cos (\theta_1 - \theta_2) + j \sin (\theta_1 - \theta_2)]$$

$$= \frac{r_1}{r_2} \underline{/\theta_1 - \theta_2}$$

Thus to divide two complex numbers divide their moduli and subtract their arguments.

For example   i $\dfrac{6\underline{/50°}}{3\underline{/26°}} = 2\underline{/24°}$

ii $\dfrac{5\underline{/45°}}{8\underline{/60°}} = \dfrac{5}{8} \underline{/-15°}$

$$= 0.625 [\cos (-15°) + j \sin (-15°)]$$
$$= 0.625 (\cos 15° - j \sin 15°)$$

12f   De Moivre's theorem

Since $(\cos \theta_1 + j \sin \theta_1)(\cos \theta_2 + j \sin \theta_2) = \cos (\theta_1 + \theta_2) + j \sin (\theta_1 + \theta_2)$

putting $\theta_1 = \theta_2 = \theta$ will give

$(\cos \theta + j \sin \theta)^2 = \cos 2\theta + j \sin 2\theta$

Multiplying both sides by $\cos \theta + j \sin \theta$

$(\cos \theta + j \sin \theta)^3 = \cos (2\theta + \theta) + j \sin (2\theta + \theta)$
$$= \cos 3\theta + j \sin 3\theta$$

Continuing this for n factors will give

$$(\cos \theta + j \sin \theta)^n = \cos n\theta + j \sin n\theta$$

This is known as De Moivre's theorem and is true not only when n is a positive integer but for all values of n, positive and negative, integral and fractional.

For example   i $(\cos 20° + j \sin 20°)^6 = \cos 120° + j \sin 120°$

$$= -\frac{1}{2} + j\frac{\sqrt{3}}{2}$$

ii $(\cos 45° + j \sin 45°)^{\frac{1}{3}} = \cos 15° + j \sin 15°$
$$= 0·966 + j0·259$$

Notice from this that $0·966 + j0·259$ is one of the cube roots of $0·707 + j707$.

12g   Applications of complex numbers

There are a number of applications of complex numbers to problems involving vector quantities. Two examples now follow, one on vector sums of forces and the other on alternating current theory.

Example 1   The following coplanar forces act away from point O.

6 N at 0°, 5 N at 30°, 3 N at 90°, 4 N at 120° and 4·5 N at 225°.

The angles given are those that the forces make with Ox. Find the resultant in magnitude and direction.

Representing the forces by complex numbers in polar form

Resultant $= 6\underline{/0°} + 5\underline{/30°} + 3\underline{/90°} + 4\underline{/120°} + 4·5\underline{/225°}$

$= 6 + 5(\cos 30° + j \sin 30°) + 3(\cos 90° + j \sin 90°)$
$\quad + 4(\cos 120° + j \sin 120°) + 4·5(\cos 225° + j \sin 225°)$
$= 6 + 4·330 + 0 - 2 - 3·182 + j(2·5 + 3 + 3·464 - 3·182)$
$= 5·148 + j5·782$
$= r\underline{/\theta}$ where $r = \sqrt{(5·148^2 + 5·782^2)} = 7·74$
$$\text{and} \quad \theta = \tan^{-1}\frac{5·782}{5·148} = 48° \ 19'$$

Resultant is 7·74 N at 48° 19' to Ox.

Example 2   A coil of resistance 5 ohms and inductance 0·02 henrys is connected in series with a capacitor of 200×10^{-6} farads across a 200 volt 50 hertz supply. Find i the impedance, ii the current taken and iii the angle of phase difference.

The current I is given by $I = \dfrac{v}{z}$ where z, the impedance of the circuit, is the vector sum of resistance R, inductive reactance $j\omega L$, and the capacitive reactance $\dfrac{1}{j\omega c} = -\dfrac{j}{\omega c}$ where $\omega = 2\pi \times$ frequency of supply.

i.e. $z = R + j\left(\omega L - \dfrac{1}{\omega c}\right)$.

i Impedance $= 5 + j\left(2\pi \times 50 \times 0.02 - \dfrac{10^6}{2\pi \times 50 \times 200}\right)$

$= 5 + j(6.28 - 15.92)$

$= 5 - j9.64$

$= r\underline{/\theta}$ where $r = \sqrt{(5^2 + 9.64^2)} = 10.86$

and $\theta = \tan^{-1} - \dfrac{9.64}{5} = -62° \ 35'$

Impedance $= 10.86\underline{/-62° \ 35'}$ ohms

ii and iii Current $= \dfrac{200}{10.86\underline{/-62° \ 35'}} = 18.4\underline{/62° \ 35'}$ A

Thus the current is 18.4 amperes and leads by $62° \ 35'$.

Exercise 12g

Express in polar form:

1 $3 + j4$ 2 -5 3 $-4 + j3$

4 $1 + j7$ 5 $-j4$ 6 $-5 + j8$

7 $(5 - j12) + (-7 - j9)$ 8 $\dfrac{5 - j2}{5 + j2}$

9 $\dfrac{(6 + j5)(3 - j4)}{(1 + j2)}$ 10 $(3 + j2)^2 + (3 - j2)^2$

Express in the cartesian form $x + jy$:

11 $4\underline{/30°}$ 12 $4.2\underline{/-50°}$ 13 $2.8\underline{/76°}$

14 $10\underline{/\frac{\pi}{2}}$ 15 $6\underline{/\pi}$ 16 $5.8\underline{/135°}$

Evaluate in polar form:

17 $3\underline{/30°} \times 5\underline{/15°}$ 18 $4.6\underline{/16°} \times 3.7\underline{/105°}$

19    $\dfrac{10\underline{/36^\circ}}{2{\cdot}5\underline{/16^\circ}}$

20    $\dfrac{3\underline{/16^\circ}}{6\underline{/-37^\circ}}$

21    $\dfrac{5\underline{/42^\circ} \times 7\underline{/112^\circ}}{6\underline{/72^\circ}}$

22    $\dfrac{3{\cdot}2\underline{/17^\circ} \times 4{\cdot}5\underline{/-60^\circ}}{16\underline{/160^\circ}}$

23   Put $3+j2$ in polar form and hence evaluate $(3+j2)^5$.

24   Evaluate $(5+j)^4$ in polar form.

25   Find the moduli and arguments of:

(a) $\cos\theta - j\sin\theta$, (b) $\sqrt{2}(1-j)$, (c) $(2+j6)-(17+j8)$

26   If $z = 1 + \cos\theta + j\sin\theta$ show that

$$|z| = 2\cos\frac{\theta}{2} \quad \text{and} \quad \arg z = \frac{\theta}{2}$$

27   Three vectors are $\mathbf{A}\ 2\underline{/30^\circ}$, $\mathbf{B}\ 3\underline{/60^\circ}$, and $\mathbf{C}\ 4\underline{/135^\circ}$. Find the vector sums (a) $\mathbf{A}+\mathbf{B}+\mathbf{C}$ and (b) $\mathbf{A}-\mathbf{B}-\mathbf{C}$, expressing the answers in the forms $x+jy$ and $r\underline{/\theta}$.

28   (a) Give the moduli and arguments of i $3+j2$, ii $(2+j6)+(17+j8)$.
(b) Find the roots of $x^2+2x+5 = 0$ in cartesian and in polar forms.
(c) Find the real and the imaginary parts of $\dfrac{1}{(1+2j)^2}$.

29   Two impedances $z_1 = 5+j18$ and $z_2 = 6-j3$ are connected in series to a supply voltage V of 250 volts. If the current I amperes is given by $I = \dfrac{V}{Z}$ where $Z = z_1+z_2$, find the current as a complex number in polar form.

30   Three circuits having impedances $z_1 = 10+j30$, $z_2 = 20+j0$, and $z_3 = 1-j20$ are connected in parallel across a 200 volt supply. If the total current is given by $\dfrac{V}{Z}$ where $\dfrac{1}{Z} = \dfrac{1}{z_1}+\dfrac{1}{z_2}+\dfrac{1}{z_3}$, find the total current flowing and its phase angle.

Miscellaneous exercises 12

1   (a) Find the modulus and argument of $(2+3j)(3-2j)$.
(b) Express in the form $x+jy$

$$\frac{1}{(2+j)^2} - \frac{1}{(2-j)^2}$$

2 (a) Express the following in the form $a+jb$ in which $j^2 = -1$ and a, b are real numbers:

i $(3-4j)(j+7)^2$,   ii $\dfrac{3+j}{2-j}$

(b) Prove that the roots x_1 and x_2 of the equation

$$x^2+2x+4 = 0 \text{ are } -1\pm j\sqrt{3}$$

If A, B, C are the points in the Argand diagram representing the numbers 2, x_1 and x_2 respectively, prove that ABC is an equilateral triangle.

E.M.E.U.

3 (a) Solve i $x^2+4 = 0$
        ii $x^2-x+1 = 0$

(b) Show that $\left(-\dfrac{1}{2}+j\dfrac{\sqrt{3}}{2}\right)^3 -1 = 0$

4 If $z_1 = 120+j50$, $z_2 = 120+j90$, $z_3 = 100-j60$, find the value of z in the cartesian form $x+jy$ if

(a) $z = z_1+z_2+z_3$ (b) $\dfrac{1}{z} = \dfrac{1}{z_1}+\dfrac{1}{z_2}+\dfrac{1}{z_3}$

5 (a) Express $Z = \dfrac{(5+3j)(2-j)}{6+7j}$ where $j^2 = -1$, in the form $a+jb$ in which a and b are real numbers, and prove that $Z^4 = -4$.

(b) O, A, B are the points in the Argand diagram representing the complex numbers 0, $4-3j$ and $7+j$, respectively. Prove that the triangle OAB is right-angled and isosceles, and find the complex number represented by the fourth vertex, C, of the square OABC. Show that this number may be written in the form

$$\dfrac{(7+j)}{\sqrt{2}}\left(\cos\dfrac{\pi}{4}+j\sin\dfrac{\pi}{4}\right)$$

6 (a) Find the roots of $x^2+x+2 = 0$ in the cartesian form $x+jy$ and in polar form $r\underline{/\theta}$.

(b) If $I = \dfrac{V}{Z}$ and $V = 100+j40$ find

i I when $Z = 20+j12$, ii Z when $I = 4+j3$.

7 (a) Obtain the roots of the equation $x^2+6x+25 = 0$ in the form $A+jB$ where $j^2 = -1$.

(b) Express $\dfrac{(3+j)(7+32j)}{1+2j}$ in the form $a+jb$, where $j^2 = -1$.

(c) Show in an Argand diagram the points A, B, C, D, which represent the complex numbers $7+5j$, $4j$, $1-3j$, $8-2j$, respectively. Prove that ABCD is a square and find the complex number which is represented by the centre of the square.

E.M.E.U.

8   If a and b are real numbers and $a-3+j(b-5) = 0$, find their values.

   $\sqrt{(3-j4)} = x+jy$. By squaring both sides and equating real and imaginary parts, show that $2-j$ is one square root of $3-j4$ and find the other.

9   Two impedances $z_1 = 5+j7$ ohms and $z_2 = 10-j5$ ohms are connected in series to a 200 volt supply. If the current I amperes is given by $I = \dfrac{V}{Z}$ where $Z = z_1+z_2$, find the current drawn from the mains as a complex number in polar form.

10 Solve the above problem if the impedances are in parallel so that

$$\frac{1}{Z} = \frac{1}{z_1}+\frac{1}{z_2}.$$

11 The following coplanar forces pass through point O, and the angles are those that the forces make with OX.

   110 mN along OX, 80 mN at 30°, 120 mN at 90°, 100 mN at 135°, 50 mN at 225°, 90 mN at 270°.

   Using complex numbers as force vectors, find the magnitude and direction of the resultant force.

12 Three vectors are $\mathbf{A}\ 3\underline{/30°}$, $\mathbf{B}\ 4\underline{/135°}$, $\mathbf{C}\ 5\underline{/330°}$. Find in polar form the vectors (a) $\mathbf{A}+\mathbf{B}+\mathbf{C}$,
      (b) $\mathbf{A}-\mathbf{B}-\mathbf{C}$.

Appendix A
Some mathematical constants and formulae

$$\pi = 3 \cdot 141\,59 \qquad\qquad \frac{1}{\pi} = 0 \cdot 318\,31$$

$$e = 2 \cdot 718\,28 \qquad\qquad \pi^2 = 9 \cdot 869\,60$$

$$\log_e 10 = 2 \cdot 302\,59 \qquad \log_{10} e = 0 \cdot 434\,29$$

$$1 \text{ radian} = \frac{180}{\pi} \text{ degrees} = 57 \cdot 296° \text{ (to 3 D)} = 57° \, 17' \, 45'' \text{ (to nearest sec)}$$

Algebra and geometry

1. $\log MN = \log M + \log N$

   $\log \dfrac{M}{N} = \log M - \log N$

   $\log N^p = p \log N$

   $\log_a N = \log_b N \times \log_a b = \dfrac{\log_b N}{\log_b a}$

2. $ax^2 + bx + c = 0$ has roots $\dfrac{-b \pm \sqrt{(b^2 - 4ac)}}{2a}$

   Roots are complex if $b^2 - 4ac$ is negative.

   Sum of roots $-\dfrac{b}{a}$

   Product of roots $\dfrac{c}{a}$

Remainder theorem

3.   If $f(x)$ is divided by $(x - p)$ the remainder is $f(p)$.

4   $_nP_r = n(n-1)\ldots(n-r+1) = \dfrac{n!}{(n-r)!}$

$_nC_r = \dbinom{n}{r} = \dfrac{n(n-1)\ldots(n-r+1)}{r!} = \dfrac{n!}{(n-r)!\,r!}$

Arithmetic progression

5   $S_n = a+(a+d)+(a+2d)+\ldots+[a+(n-1)d]$

$= \dfrac{n}{2}[2a+(n-1)d]$

Geometric progression

$S_n = a+ar+ar^2+\ldots+ar^{n-1}$

$= \dfrac{a(r^n-1)}{r-1}$

$S_\infty = \dfrac{a}{1-r} \qquad |r| < 1$

Binomial theorem

6   (a) n positive integer $(a+x)^n = a^n+\dbinom{n}{1}a^{n-1}x+\ldots\dbinom{n}{r}a^{n-r}x^r+\ldots+x^n$

(b) n fractional or negative $(1+x)^n = 1+\dbinom{n}{1}x+\dbinom{n}{2}x^2\ldots+\dbinom{n}{r}x^r+\ldots$
$|x| < 1$

7       $e^x = 1+x+\dfrac{x^2}{2!}+\dfrac{x^3}{3!}+\ldots$ valid for all values of x

$e^{-x} = 1-x+\dfrac{x^2}{2!}-\dfrac{x^3}{3!}+\ldots$ valid for all values of x

$\log_e(1+x) = x-\dfrac{x^2}{2}+\dfrac{x^3}{3}-\ldots$ valid for $|x| < 1$

$\log_e(1-x) = -x-\dfrac{x^2}{2}-\dfrac{x^3}{3}-\ldots$ valid for $|x| < 1$

$\sin x = x-\dfrac{x^3}{3!}+\dfrac{x^5}{5!}-\ldots x$ in radians

$\cos x = 1-\dfrac{x^2}{2!}+\dfrac{x^4}{4!}-\ldots x$ in radians

Equations of a straight line

8
$$y = mx + c$$

$m =$ gradient

$c =$ intercept on y-axis

$$y - y_1 = m(x - x_1)$$

(x_1, y_1) point on line

Two lines are (a) parallel if $m_1 = m_2$

(b) perpendicular if $m_1 m_2 = -1$

Equation of a circle

9 Centre at origin, radius a, $x^2 + y^2 = a^2$.

Newton's approximation

10 If a is an approximate solution of $f(x) = 0$ then $a - \dfrac{f(a)}{f'(a)}$ is a better approximation.

Complex numbers

11
$$z = x + jy = r(\cos\theta + j\sin\theta) = r\underline{/\theta}$$

$$z_1 z_2 = r_1\underline{/\theta_1} \times r_2\underline{/\theta_2} = r_1 r_2\underline{/\theta_1 + \theta_2}$$

$$\frac{z_1}{z_2} = \frac{r_1\underline{/\theta_1}}{r_2\underline{/\theta_2}} = \frac{r_1}{r_2}\underline{/\theta_1 - \theta_2}$$

$(\cos\theta + j\sin\theta)^n = \cos n\theta + j\sin n\theta$ true for all values of n.

Trigonometry and mensuration

1
$$\sec A = \frac{1}{\cos A}; \operatorname{cosec} A = \frac{1}{\sin A}; \cot A = \frac{1}{\tan A}$$

$$\frac{\sin A}{\cos A} = \tan A; \frac{\cos A}{\sin A} = \cot A$$

2 $\sin(-\theta) = -\sin\theta$, $\cos(-\theta) = \cos\theta$, $\tan(-\theta) = -\tan\theta$,

$\sin(90 - \theta) = \cos\theta$, $\cos(90 - \theta) = \sin\theta$, $\tan(90 - \theta) = \cot\theta$

3 $\sin^2 A + \cos^2 A = 1$, $1 + \tan^2 A = \sec^2 A$, $1 + \cot^2 A = \operatorname{cosec}^2 A$

4 $\sin(A \pm B) = \sin A \cos B \pm \cos A \sin B$

$\cos(A \pm B) = \cos A \cos B \mp \sin A \sin B$

$$\tan(A \pm B) = \frac{\tan A \pm \tan B}{1 \mp \tan A \tan B}$$

5 $\sin 2A = 2 \sin A \cos A$

$\cos 2A = \cos^2 A - \sin^2 A = 2 \cos^2 A - 1 = 1 - 2 \sin^2 A$

$\tan 2A = \dfrac{2 \tan A}{1 - \tan^2 A}$

6 $2 \sin A \cos B = \sin (A + B) + \sin (A - B)$
$2 \cos A \sin B = \sin (A + B) - \sin (A - B)$
$2 \cos A \cos B = \cos (A + B) + \cos (A - B)$
$2 \sin A \sin B = \cos (A - B) - \cos (A + B)$

7 $\sin A + \sin B = 2 \sin \dfrac{A + B}{2} \cos \dfrac{A - B}{2}$

$\sin A - \sin B = 2 \cos \dfrac{A + B}{2} \sin \dfrac{A - B}{2}$

$\cos A + \cos B = 2 \cos \dfrac{A + B}{2} \cos \dfrac{A - B}{2}$

$\cos A - \cos B = -2 \sin \dfrac{A + B}{2} \sin \dfrac{A - B}{2}$

8 $a \sin \theta + b \cos \theta = \sqrt{(a^2 + b^2)} \sin \left(\theta + \tan^{-1} \dfrac{b}{a} \right)$

$= \sqrt{(a^2 + b^2)} \cos \left(\theta - \tan^{-1} \dfrac{a}{b} \right)$

9 If $\sin \theta = \sin \alpha, \quad \theta = n\pi + (-1)^n \alpha$
$\cos \theta = \cos \alpha, \quad \theta = 2n\pi \pm \alpha$
$\tan \theta = \tan \alpha, \quad \theta = n\pi + \alpha$

10 $\dfrac{a}{\sin A} = \dfrac{b}{\sin B} = \dfrac{c}{\sin C}$

$a^2 = b^2 + c^2 - 2bc \cos A$

$\tan \dfrac{B - C}{2} = \dfrac{b - c}{b + c} \cot \dfrac{A}{2}$

$\sin \dfrac{A}{2} = \sqrt{\dfrac{(s - b)(s - c)}{bc}} \qquad \cos \dfrac{A}{2} = \sqrt{\dfrac{s(s - a)}{bc}}$

Area $= \frac{1}{2}ab \sin C = \sqrt{[s(s - a)(s - b)(s - c)]}$

Areas

11 Circle πr^2
Sector $\frac{1}{2}r^2\theta = \frac{1}{2}rs$ $(s = \text{arc} = r\theta)$

Segment $\frac{1}{2}r^2(\theta - \sin\theta)$
Cylinder: curved surface $2\pi rh$
Cone: curved surface πrl (l = slant height)
Sphere: curved surface $4\pi r^2$
Zone or cap: curved surface $2\pi rt$

Irregular areas

12   Mid-ordinate rule $h(y_1 + y_2 + \ldots + y_n)$

Trapezoidal rule $h\left(\dfrac{y_1 + y_{n+1}}{2} + y_2 + \ldots + y_n\right)$

Simpson's rule $\dfrac{h}{3}(A + 2B + 4C)$

(Even number of strips) $A =$ first + last
                            $B =$ remaining odd ordinates
                            $C =$ even ordinates

Volumes

13   Cylinder $\pi r^2 h$
Cone $\frac{1}{3}\pi r^2 h$
Sphere $\frac{4}{3}\pi r^3$
Pyramid $\frac{1}{3}$ base $\times$ height

Truncated pyramid and cone $\dfrac{h}{3}(A_1 + \sqrt{(A_1 A_2)} + A_2)$

Segment of sphere $\dfrac{\pi h}{6}(3a^2 + h^2) = \pi h^2\left(r - \dfrac{h}{3}\right)$

Calculus

Differential coefficients

1

y	$\dfrac{dy}{dx}$
x^n	nx^{n-1}
$\sin x$	$\cos x$
$\cos x$	$-\sin x$
$\tan x$	$\sec^2 x$
$\sec x$	$\sec x \tan x$
$\operatorname{cosec} x$	$-\operatorname{cosec} x \cot x$
$\cot x$	$-\operatorname{cosec}^2 x$
e^x	e^x
a^x	$a^x \log_e a$
$\log_e x$	$\dfrac{1}{x}$

y	$\dfrac{dy}{dx}$
uv	$u\dfrac{dv}{dx}+v\dfrac{du}{dx}$
$\dfrac{u}{v}$	$\dfrac{1}{v^2}\left(v\dfrac{du}{dx}-u\dfrac{dv}{dx}\right)$
$y=f(u),\ u=g(x)$	$\dfrac{dy}{dx}=\dfrac{dy}{du}\times\dfrac{du}{dx}$

Approximations

2 (a) y a function of x only

$$\delta y \simeq \frac{dy}{dx}\times\delta x$$

(b) V a function of x, y, z

$$\delta V \simeq \frac{\partial V}{\partial x}\,\delta x+\frac{\partial V}{\partial y}\,\delta y+\frac{\partial V}{\partial z}\,\delta z$$

Maximum and minimum values for $y = f(x)$

3 (a) Put $\dfrac{dy}{dx}=0$ and solve for x.

(b) Test sign of $\dfrac{d^2y}{dx^2}$ ($-$ for maximum, $+$ for minimum)

Standard integrals

4	y	$\int y\,dx$
	x^n	$\dfrac{x^{n+1}}{n+1}+c \quad n\neq-1$
	$\dfrac{1}{x}$	$\log x+c$
	$\cos x$	$\sin x+c$
	$\sin x$	$-\cos x+c$
	$\sec^2 x$	$\tan x+c$
	$\tan x$	$\log\sec x+c$
	e^x	e^x+c
	a^x	$\dfrac{a^x}{\log_e a}+c$

(a) The addition of a constant to x makes no difference in the form of the integral.

e.g. $\int (x+7)^n \, dx = \dfrac{(x+7)^{n+1}}{n+1} + c$

and $\displaystyle\int \dfrac{1}{x+2} \, dx = \log(x+2) + c$

(b) If x is multiplied by a constant, the integral is of the same form, but is divided by the constant.

e.g. $\int (5x+7)^n \, dx = \dfrac{(5x+7)^{n+1}}{5(n+1)} + c$

$\int e^{3x} \, dx = \dfrac{e^{3x}}{3} + c$

Integration by parts

5    $\int u \, dv = uv - \int v \, du$

Applications of integration

6   For the curve $y = f(x)$ between limits $x = a$, $x = b$, and x-axis.

(a) Area $A = \displaystyle\int_a^b y \, dx$

(b) C.G. of area $\bar{x} = \dfrac{1}{A} \displaystyle\int_a^b xy \, dx$; $\bar{y} = \dfrac{1}{A} \displaystyle\int_a^b \dfrac{y^2}{2} \, dx$

(c) $k_{OX}^2 = \dfrac{1}{A} \displaystyle\int_a^b \dfrac{y^3}{3} \, dx$

$k_{OY}^2 = \dfrac{1}{A} \displaystyle\int_a^b x^2 y \, dx$

(d) Mean value $\dfrac{1}{b-a} \displaystyle\int_a^b y \, dx$

(e) R.M.S. value $\sqrt{\left(\dfrac{1}{b-a} \displaystyle\int_a^b y^2 \, dx \right)}$

(f) Volume of solid of revolution about OX, $\pi \int\limits_a^b y^2 \, dx$

(g) For solid of revolution about OX

$$I_{OX} = \frac{\pi}{2} \int\limits_a^b \rho y^4 \, dx; \; I_{OY} = \pi \int\limits_a^b \rho y^2 \left(\frac{y^2}{4} + x^2\right) dx$$

7 (a) $I_{OZ} = I_{OX} + I_{OY}$ OX, OY in plane of lamina.
   $k_{OZ}^2 = k_{OX}^2 + k_{OY}^2$
   (b) $I_H = I_G + Mh^2$ h distance between axes.
   $k_H^2 = k_G^2 + h^2$

Differential equations

8 (a) $f(y) \dfrac{dy}{dx} = g(x)$ has solution $\int f(y) \, dy = \int g(x) \, dx + c$

(b) $\dfrac{d^2 y}{dx^2} + n^2 y = 0$ has solution $y = A \sin nx + B \cos nx$ or $y = a \sin(nx + \alpha)$

Statistics

Mean $\bar{x} = \dfrac{1}{N} \sum\limits_{r=1}^{r=n} f_r x_r$

Variance $\sigma^2 = \dfrac{1}{N} \sum\limits_{r=1}^{r=n} f_r(x_r - \bar{x})^2$

Second moment about m $\mu^2 = \sigma^2 + (\bar{x} - m)^2$

Mean deviation from mean $= \dfrac{1}{N} \sum\limits_{r=1}^{r=n} f_r|(x_r - \bar{x})|$

Coefficient of variation $\dfrac{\sigma}{\bar{x}} \times 100\%$

Standard error of the mean $\dfrac{\sigma}{\sqrt{n}}$

Probability (r times in n trials)

$P(r) = \dbinom{n}{r} p^r q^{n-r}$ (binomial)

$\simeq \dfrac{e^{-m} m^r}{r!}$ (Poisson: $m = np$; p small)

Binomial distribution $\bar{x} = np$, $\sigma = \sqrt{(npq)}$
Poisson distribution $\bar{x} = np$, $\sigma = \sqrt{(np)}$

Interpolation formula

$$f_p = f_0 + p\Delta f_0 + \frac{p(p-1)}{2!}\Delta^2 f_0 + \ldots$$

Système International

Basic SI units with derived units

Quantity	Name of unit	Symbol
length	metre	m
mass	kilogramme	kg
time	second	s
electric current	ampere	A
absolute temperature	kelvin	K
luminous intensity	candela	cd
plane angle	radian	rad
force	newton	$N = kg\,m\,s^{-2}$
work, energy	joule	$J = N\,m$
power	watt	$W = J\,s^{-1}$
electric potential	volt	$V = W\,A^{-1}$
electric resistance	ohm	$\Omega = V\,A^{-1}$
inductance	henry	$H = V\,s\,A^{-1}$
frequency in cycles per second	hertz	$Hz = s^{-1}$

Prefixes denoting multiples and submultiples

Factor	Prefix	Symbol
10^{12}	tera-	T
10^{9}	giga-	G
10^{6}	mega-	M
10^{3}	kilo-	k
10^{2}	hecto-	h
10	deca-	da
10^{-1}	deci-	d
10^{-2}	centi-	c
10^{-3}	milli-	m
10^{-6}	micro-	μ
10^{-9}	nano-	n
10^{-12}	pico-	p
10^{-15}	femto-	f
10^{-18}	atto-	a

U.K. units in terms of SI units

length   1 ft = 0·3048 m

                 1 in = 0·0254 m

mass   1 lb = 0·4536 kg

density   1 lb ft^{-3} = 16·02 kg m^{-3}

force   1 lbf = 4·448 N

pressure   1 lbf in^{-2} = 6895 N m^{-2}

energy   1 ft lbf = 1·356 J

               1 Btu = 1055 J

power   1 hp = 745·7 W

Appendix B
Ordinary National Certificate mathematics examination papers

Examination paper 1

Time allowed: 3 hours. Answer SIX questions.
A calculating machine must be used in question 4.

1 (a) If p and q are the roots of the equation $3x^2 + x + 1 = 0$ find the values of $p + q$, pq, and $\frac{1}{p} + \frac{1}{q}$ without solving the equation.

(b) Solve the equation $4^{x+1} + 3 \times 2^x - 1 = 0$.

(c) Resolve $\dfrac{3x^2 - 2x + 1}{(x-2)(x+1)^2}$ into three partial fractions.

2 (a) Show that if x^3 and higher powers of x can be neglected

$$\frac{\sqrt{(1+x)}}{(1-x)^2} = 1 + \frac{5}{2}x + \frac{31}{8}x^2$$

Use this expansion to evaluate $\dfrac{\sqrt{1\cdot02}}{(0\cdot98)^2}$.

(b) In the formula $h = \dfrac{klv^2}{d}$ find the approximate percentage error in h due to an error of $2\frac{1}{2}\%$ in l, -2% error in v, and $+1\frac{1}{2}\%$ error in d.

3 (a) Express $5 \sin \omega t - 12 \cos \omega t$ in the form $R \sin (\omega t - \alpha)$. Find the maximum value of the expression, and the smallest positive value of t at which it occurs.

(b) Solve the equation $\sin \theta - 2 \cos 2\theta = 1$ giving all values of θ from $0°$ to $360°$.

(c) Show that $\sin x + \sin \left(x + \dfrac{2\pi}{3}\right) + \sin \left(x + \dfrac{4\pi}{3}\right) = 0$.

4 (a) Solve the simultaneous equations to find P and W correct to four significant figures.

$2\cdot315\,P - 4\cdot536\,W + 2\cdot781 = 0$
$3\cdot763\,P + 1\cdot281\,W - 6\cdot437 = 0$

(b) Solve the equation $x^2 + 2x - 1 = 0$ by rearranging it in the form $x = \dfrac{1}{x+2}$ and using an iterative method. Give your solutions correct to five decimal places.

5 (a) Differentiate with respect to x

   i $e^{2x} \sin 3x$ ii $\dfrac{\log{(2x+1)}}{x^2}$ iii $\sqrt{(x^2 - 1)}$

   (b) Find the maximum and minimum values of $y = x + \dfrac{1}{x}$. Sketch the graph of the function.

6 (a) Obtain the following integrals

   i $\displaystyle\int \dfrac{(x^2-1)(x^2+1)}{x^2}\,dx$ ii $\int 3e^{\frac{1}{2}x}\,dx$ iii $\displaystyle\int_{0}^{\frac{1}{2}\pi} (\sin x + \cos 2x)\,dx$

   (b) Find i by integration ii by Simpson's rule, the area between the curve $y = \dfrac{10}{x^2}$, the x-axis and the ordinates $x = 2$ and $x = 8$. (Use six strips in the second method and work to four decimal places.)

7 (a) By solving the differential equation $\dfrac{dy}{dx} = \dfrac{\sin x}{y^8}$ show that

   $y^9 + 9 \cos x + 8 = 0$ if $y = 1$ when $x = \pi$.

   (b) The velocity v m s^{-1} of a particle is given by $v = \frac{1}{2}t^2 + 4t^{\frac{3}{2}} - 3t$ where t is the time in seconds.
   i Find the acceleration of the particle when $t = 4$.
   ii Find the displacement, x m, of the particle after nine seconds if $x = 0$ when $t = 0$.

8 (a) Show that the second moment of area of an annulus, of inner and outer radii R_1 and R_2 respectively, about an axis through its centroid perpendicular to its plane is $\dfrac{\pi}{2}(R_2^4 - R_1^4)$ and hence deduce its radius of gyration about this axis.
   (b) Using the theorems of perpendicular and parallel axes obtain the second moment of area of the same annulus i about a diameter, ii about an axis parallel to a diameter and tangential to the outer edge.

9 The breaking loads in kilonewtons of 100 steel specimens are given in the following frequency table.

Centre of interval	32·0	32·2	32·4	32·6	32·8	33·0	33·2	33·4	33·6	33·8
Frequency	1	4	11	13	21	24	12	9	3	2

Find the mean and the standard deviation of the distribution. Construct a cumulative frequency table and draw the ogive.

10   (a) Write down the series for $\log_e (1+x)$ and for $\sin x$. Hence expand

$$\log_e (1+x) \sin x$$

as a series in ascending powers of x as far as the term in x^4.

(b) If $V = \frac{1}{3}\pi r^2 h$ show that

$$\frac{\delta V}{V} \simeq \frac{2\delta r}{r} + \frac{\delta h}{h}$$

Hence find the approximate percentage change in the volume of a right circular cone if the radius increases from 5 to 5·03 and the height decreases from 10 to 9·95.

Examination paper 2

Time allowed: 3 hours. Answer SIX *questions.*
A calculating machine must be used in question 6 (a).

1   (a) Write down the first four terms of the expansion $(1-x^2)^{-2}$ and hence find

a value of $\dfrac{1}{(1·004)^2}$ correct to four decimal places.

(b) Quote the series for e^x and expand $\sqrt{e^x} + \dfrac{1}{\sqrt{e^x}}$ as far as the term in x^4.

(c) In the formula $W = \dfrac{Cbd^3}{L}$ where C is a constant, find the approximate percentage change in W if b increases by 2%, d decreases by $1\frac{1}{2}\%$ and L increases by 3%.

2   (a) Express $\cos A - \cos 5A$ and $\sin 5A - \sin A$ each as a product and hence show that

$$\frac{\cos A - \cos 5A}{\sin 5A - \sin A} = \tan 3A$$

(b) Express $4 \sin \theta - 7 \cos \theta$ in the form $R \sin (\theta - \alpha)$ giving the values of R and α. Hence find the maximum value of $4 \sin \theta - 7 \cos \theta$ and the value of θ at which the maximum occurs.

3 (a) Differentiate with respect to x

   i $\dfrac{1+x^2}{1-x}$ ii $\sin 3x \cos 2x$ iii $\log_e \sqrt{(x^2-3)}$

(b) Water is running into an inverted cone of semi-vertical angle $45°$ at a steady rate of $4 \text{ cm}^3 \text{ s}^{-1}$. Find the rate at which the level is rising when the depth of water is i 2 cm ii 5 cm.

4 (a) Rearrange the following in a straight line law of the form $Y = aX+b$, where X and Y are functions of x and y.

   $$y = ax^2+b; \; y = \dfrac{1}{ax+b}; \; y = ab^x$$

(b) The following values of t and R are believed to obey a law of the form $R = An^t$.

t	0·2	0·6	1·0	1·5	2·0	2·5	3·0
R	7·2	15	34	85	230	620	1560

State the most suitable type of logarithmic ruled graph paper to use when testing the relationship. Show that the values are so related and find suitable values for A and n.

5 The following table gives the lengths of 600 bolts measured to the nearest 0·01 cm. Calculate the mean length. Plot a cumulative frequency graph and find the median, the upper and lower quartiles and the semi-interquartile range.

Length (cm)	4·95	4·96	4·97	4·98	4·99	5·00	5·01	5·02	5·03	5·04	5·05	Total
Frequency	3	19	63	120	173	126	70	15	6	3	2	600

6 (a) Using an iterative method evaluate $\sqrt{21}$ correct to five significant figures.
  (b) The following table gives the angle θ turned through by a shaft for equal values of the time t.

t (s)	0	0·2	0·4	0·6	0·8	1·0	1·2
θ (rad)	0	1·52	3·24	5·12	7·22	9·56	12·14

Construct a finite difference table up to and including the fourth order. Using the Gregory–Newton interpolation formula estimate the value of θ when $t = 0·52$ s.

7 (a) Obtain the integrals

   i $\displaystyle\int \left(3\sqrt{x} - \dfrac{3}{x^2}\right) dx$ ii $\int \sin^2 x \cos x \, dx$

(b) Evaluate i $\int_1^e x^2 \log_e x \, dx$ ii $\int_0^3 \frac{dx}{(x+3)(x+5)}$

8   The curves $y^2 = 4x$ and $xy = 2$ intersect in the first quadrant. Find the area enclosed between the curves from their point of intersection to the ordinate at $x = 4$. Find also the volume obtained by the rotation of this area about the x-axis. Hence, without further integration, find the distance of the centroid of the area from the x-axis.

9   (a) If $z = \sin(2x+3y)$ find $\dfrac{\partial z}{\partial x}$ and $\dfrac{\partial z}{\partial y}$.

(b) Show that $2x^3 - 4x + 1 = 0$ has three roots between -2 and $+2$. Use Newton's method to find the largest root correct to two decimal places.

10   (a) Write down the equation of the straight line of gradient m which passes through the point (x_1, y_1). Find the equation of the tangent to the circle $x^2 + y^2 = 20$ at the point $(4, 2)$.

(b) Solve the differential equation

$$(2x-1)\frac{dy}{dx} = y+1 \quad \text{given that } y = 1 \text{ when } x = 1.$$

Examination paper 3

Time allowed: 3 hours. Answer SIX *questions.*
A calculating machine must be used in question 6 (b).

1   (a) Find x if $5e^{2x} - 12e^x + 4 = 0$.

(b) If the roots of $2x^2 + x + 1 = 0$ are α and β find the values of $\alpha\beta$, $\alpha+\beta$, $\alpha^2 + \beta^2$ without solving the equation.

(c) Resolve $\dfrac{x^2 - 5x + 3}{(x+2)(x-1)(x-2)}$ into three partial fractions.

2   (a) If θ is an acute angle and $\sin\theta = 0{\cdot}6$ find, without using tables, the value of $\sin 2\theta + 3\cos 2\theta$.

(b) Solve the equation $\operatorname{cosec}\theta - 6\sin\theta = 1$ giving all the solutions between $0°$ and $360°$.

(c) Show that $\dfrac{\sin A - \sin 3A + \sin 5A}{\cos A - \cos 3A + \cos 5A} = \tan 3A$.

3   (a) Find the first four terms in the binomial expansions of

i $(1+x)^{\frac{1}{3}}$ and   ii $(1-2x)^{-\frac{1}{3}}$

Hence expand $\sqrt[3]{\dfrac{1+x}{1-2x}}$ as far as the term in x^3.

(b) Write down the series for $\log_e (1+x)$ and $\log_e (1-x)$.

If $e^{0.04t} = \dfrac{1-x}{1-3x}$ when $0 < x < \frac{1}{3}$

express t as a series in ascending powers of x as far as the term in x^4.

4 (a) Differentiate with respect to x

    i $e^{2x}(x^2-3x)$ ii $\dfrac{x}{x^2+1}$ iii $\tan^2 (2x-1)$

(b) Find the coordinates of the points on the curve $y = x^3 + \frac{1}{2}x^2 + 6$ at each of which the tangent to the curve is parallel to the straight line $y = 2x + 5$.

5 (a) Find the roots of the equation $x^2 + x + 3 = 0$ in the form $a + jb$ where $j^2 = -1$.

(b) Express $\dfrac{(2+j)(5-j6)}{1+j2}$ in the form $a + jb$.

(c) If $I = \dfrac{V}{z_1 + z_2}$ where $V = 200$, $z_1 = 5 + j8$, and $z_2 = 7 - j3$, find I in the polar form $r \underline{/\theta}$.

6 (a) Solve the equation $\log_{10} x + x = 4$ by an iterative method giving your answer correct to four significant figures.

(b) From the following table evaluate $\int_0^1 f(x)\, dx$ to three decimal places.

x	0	0·1	0·2	0·3	0·4	0·5	0·6	0·7	0·8	0·9	1·0
$f(x)$	2	1·9802	1·9231	1·8349	1·7241	1·6000	1·4706	1·3423	1·4195	1·1050	1

7 (a) Obtain i $\displaystyle\int \dfrac{(x-1)^2}{x}\, dx$ ii $\displaystyle\int \dfrac{dx}{(2+3x)^2}$

(b) Evaluate $\displaystyle\int_0^{\frac{1}{3}\pi} \cos (3x+1)\, dx$.

(c) Find the root mean square value of $y = 4 \sin x$ over the range $x = 0$ to $x = \pi$.

8 A cap of a sphere is generated by the revolution about the x-axis of the area under the curve $4x^2 + 4y^2 = 25$ between the ordinates $x = \frac{3}{2}$ and $x = \frac{5}{2}$.

(a) Show that the volume of the cap is $\dfrac{13\pi}{6}$.

(b) Find the position of the centre of gravity of the cap.

9   Show that $x = a \sin (nt + c)$ is a solution of the differential equation

$$\frac{d^2x}{dt^2} + n^2x = 0$$

Find the solution of the differential equation

$$\frac{d^2x}{dt^2} + 16x = 0 \quad \text{given that } x = 12 \text{ and } \frac{dx}{dt} = 0 \text{ when } t = 0.$$

What is the amplitude and frequency of the oscillations?

10 (a) A normal distribution has a mean of 60 and a standard deviation of 4. Find the area under the distribution curve from $x = 56$ to $x = 66$.

(b) A certain make of light bulb has an average life of 1600 hours and a standard deviation of 60 hours. Assuming a normal distribution, what number of bulbs from a batch of 200 can be expected to fail after 1510 hours? How many should last at least 1630 hours?

Appendix C
Answers

1b

1. i $-3, -1$; real ii $-1, 4$; complex iii $-\frac{2}{3}, -\frac{1}{3}$; real
2. 11
3. i $x^2 - 2x - 3 = 0$ ii $2x^2 - x - 1 = 0$ iii $x^2 - 3kx + 2k^2 = 0$
4. $k = 9$ or 1; roots -3 or 1
5. $-\frac{5}{2}, 2, 2\frac{1}{4}$; $8x^2 - 9x + 8 = 0$
7. i $1, -1, 2, -2$ ii $\pm\frac{1}{3}, \pm 1$
8. i $1, -2, \dfrac{-1 \pm \sqrt{19}}{2}$ ii $\dfrac{3 \pm \sqrt{(-7)}}{4}, \dfrac{-1 \pm \sqrt{(-3)}}{2}$
9. i 1 ii $1, 0$ iii $-3\cdot37$
10. $c = -2\frac{7}{8}$

1c

1. $\dfrac{2}{x-2} - \dfrac{1}{x-3}$

2. $\dfrac{1}{1-x} - \dfrac{3}{2-x}$

3. $\dfrac{1}{x+2} + \dfrac{1}{x+3}$

4. $\dfrac{1}{2(x+1)} - \dfrac{1}{2(x+3)}$

5. $\dfrac{1}{a-b}\left(\dfrac{1}{x-a} - \dfrac{1}{x-b}\right)$

6. $\dfrac{1}{x+1} - \dfrac{2}{x+2} + \dfrac{1}{x+3}$

7. $\dfrac{2}{x+1} + \dfrac{2}{x-1}$

8. $\dfrac{3}{(x-2)^2} - \dfrac{1}{x-2}$

9. $\dfrac{2}{x-2} + \dfrac{3}{x+1} - \dfrac{5}{(x+1)^2}$

10. $\dfrac{7}{4(x-3)} + \dfrac{9}{4(x+1)} - \dfrac{2}{(x+1)^2}$

11. $\dfrac{1}{x} - \dfrac{2}{(x+1)^2}$

12. $\dfrac{3}{x-1} + \dfrac{1-3x}{1+x^2}$

13. $\dfrac{7(x+1)}{3(x^2+2x-2)} - \dfrac{4}{3(x+1)}$

14. $x + \dfrac{4}{3(x-1)} - \dfrac{1}{3(x+2)}$

15. $1 + \dfrac{1}{x} + \dfrac{2x+4}{x^2+x+1}$

16. $\dfrac{1}{x-1} - \dfrac{1}{x+1} - \dfrac{2}{x^2}$

Miscellaneous exercises 1

1. (a) 1.586 (b) $\dfrac{1}{x-3}+\dfrac{3}{x+1}-\dfrac{2}{(x+1)^2}$ (c) $-2\frac{1}{2}$

2. (a) $-3, -2$; real (b) $-2, 2$; complex (c) $\frac{3}{4}, -\frac{1}{4}$; real

3. $b^2 \geq 4ac$; b opposite sign to a and c; $m = 12$

4. (a) 2 or -3 (b) $a = 2\frac{1}{3}; 1\frac{1}{3}$

5. (a) $-4, 3\frac{1}{2}, 9$ (b) -2

6. $\frac{1}{2}, 1\frac{1}{2}, -\frac{3}{4}; 8x^2+2x-3 = 0$

7. (a) $\pm 2, \pm\sqrt{2}$ (b) $\pm 1, \pm 2$

8. (a) 11 (b) -1.99 (c) $25, 1$

9. (a) $\dfrac{4}{3(x-2)}+\dfrac{5}{3(x+1)}$ (b) $\dfrac{2}{1+2x}+\dfrac{3}{(1-x)^2}-\dfrac{1}{1-x}$

10. $\dfrac{3}{x-1}-\dfrac{1}{(x-1)^2}+\dfrac{2}{2x+1}$

11. $\dfrac{1}{x-3}+\dfrac{2}{x+2}-\dfrac{4}{(x+2)^2}$

12. (a) $k = 1$ or 2 (b) $\dfrac{2}{x-1}-\dfrac{3}{x+2}+\dfrac{2x+1}{x^2+1}$

13. (a) $3, 10$ (b) 1.00

14. (a) $-0.693, -1.10$ (b) $2.73, -0.732$ (c) $2, -3$

15. (a) $-\frac{2}{3}$ (b) $0, 1.26$ (c) $\dfrac{2}{x+2}+\dfrac{3}{x-1}-\dfrac{1}{(x-1)^2}$

16. (a) 13 (b) 2 (c) $(2, 4) (-2, -4), (4, 2) (-4, -2)$

17. $1.00\ \mathrm{GN\ m^{-2}}$; $988\ \mathrm{MN\ m^{-2}}$ compression

2a 1. 3.1623 2. 4.1231 3. (a) $0.4142, -2.4142$
    4. 2.571 (b) $0.3028, -3.3028$

2d 1. $2x^2-3x-2$ 2. $f(1) = 1$ 3. x^3-2x^2-4x+3
    4. $0.1674, 0.1593$ 5. $1.9796, 1.9845$
    6. $270.9, 312.6$ 7. $0.4020, 0.3924$
    8. At $x = 5, f(x) = 3.25$ (the settled value of $\Delta^3 f = 12$)

Miscellaneous exercises 2

1. (a) 4.3589 (b) $-0.4142, 2.4142$
3. $\frac{23}{28}$
4. (a) $3.466, 3.471$ (b) $2.080\ 30, 2.088\ 43$
5. $-43, 18.0, 6.3$
6. $f(1.2) = 5.256$

7. $a = 2, b = -3, c = 1$
8. (a) 1154·1 (b) 1154·2
9. 1·6737, 1·6550
10. 1·43438
11. $s = t^2 - 6t + 6$; 4 cm s^{-1}, 2 cm s^{-2}
12. (a) 3·2867 ± 0·0001 (b) 0·119 889 7

3a 2. $\dfrac{24}{25}, -\dfrac{7}{25}, -\dfrac{24}{7}$

3. $\dfrac{\sqrt{3}+1}{2\sqrt{2}}, \dfrac{\sqrt{3}-1}{2\sqrt{2}}, \dfrac{\sqrt{3}+1}{2\sqrt{2}}$

4. $\dfrac{56}{65}, \dfrac{16}{63}$

11. $44° 43', 104° 13', 31° 4'$

12. $45° 44', 32° 14', 102° 2'$

3b
1. $\cos 8\theta + \cos 2\theta$
2. $\cos 2\theta - \cos 8\theta$
3. $\sin 7\theta - \sin \theta$
4. $\sin 11\theta + \sin 3\theta$
5. $\cos 8\theta + \cos 14\theta$
6. $\cos 28° - \cos 80°$
7. $\frac{1}{2} - \frac{1}{2} \sin 40°$
8. $\frac{1}{2} \cos 40° - \frac{1}{4}$
9. $\frac{1}{2} \sin 70° - \frac{1}{4}$
10. $\frac{1}{2} \sin 6x + \frac{1}{2} \sin 4x$
11. $\frac{1}{2} \cos 2x + \frac{1}{2} \cos 4x$
12. $\frac{1}{2} \cos 2x - \frac{1}{2} \cos 6x$
13. $\sin (2x + 3y) - \sin y$
14. $\cos 2y - \cos 2x$

3c
1. $2 \cos 45° \sin 30°$
2. $2 \cos 45° \cos 30°$
3. $-2 \sin 40° \sin 12°$
4. $2 \sin 30° \cos 20°$
5. $2 \sin 28° \cos 4°$
6. $-2 \sin 60° \sin 20°$
7. $2 \sin 5x \cos x$
8. $2 \cos 5x \cos x$
9. $2 \cos 5x \sin x$
10. $-2 \sin 5x \sin x$
11. $2 \cos 6x \sin x$
12. $2 \cos 2x \cos x$
21. $a = 186, B = 118° 12', C = 39° 32'$
22. $a = 42·1, B = 107° 6', C = 44° 22'$

3d
1. (a) 150, $\dfrac{\pi}{10}, \dfrac{10}{\pi}$, 0 (b) 200, $\dfrac{1}{50}$, 50, 0 (c) $y_0, \dfrac{2\pi b}{a}, \dfrac{a}{2\pi b}, \dfrac{bc}{a}$

(d) 4·2, $\dfrac{\pi}{30}, \dfrac{30}{\pi}, \dfrac{1}{3000}$ (e) 0·17, $\dfrac{\pi}{5}, \dfrac{5}{\pi}$, -0·15

2. (a) $100 \sin 120\pi t$ (b) $100 \cos 120\pi t$ (c) $100 \sin \left(120\pi t + \dfrac{\pi}{6}\right)$

3. 3·7, 0·151, 1·66, 0·0385, 0·604
4. $25 \sin (20\pi t + 1·287)$; 25, 10, $1·287^c = 73° 44'$
5. $100 \sin (100\pi t + 0·644)$, 100, $\frac{1}{50}$, 50, $0·644^c$

Miscellaneous exercises 3

1. (a) $20\sqrt{3}\cos\theta$ (b) $3(\cos 4x + \cos 14x)$
2. (a) $2\cos 42°\cos 6°$, $2\cos 40°\sin 10°$
   (b) $\frac{1}{2}(\sin 70° - \sin 30°)$; $\frac{1}{2}(\cos 2x - \cos 8x)$
   (c) $2\cdot4$, $1\cdot80$; $0\cdot9$, $1\cdot80$; $2\cdot2$
3. (b) $20°$, $22\frac{1}{2}°$, $67\frac{1}{2}°$, $100°$, $112\frac{1}{2}°$, $140°$, $157\frac{1}{2}°$
4. $10\cdot4\sin(2t+12°\,9')$; $0\cdot318$; $10\cdot4$
5. (b) $142°\,42'$
6. $6\cdot84\sin(3t+0\cdot910)$; $6\cdot84$, $2\cdot09$ s; $0\cdot536$ s

7. (a) $60\sin 80\pi t$ (b) $60\cos 80\pi t$ (c) $60\sin\left(80\pi t + \dfrac{\pi}{6}\right)$

8. (b) $1\cdot47$, $2\cdot21$, $4\cdot07$, $4\cdot81$
9. (a) 4
10. (a) i $\frac{1}{4}(\sin 2x + \sin 4x - \sin 6x)$ ii $\frac{1}{2}\sin 2x + 0\cdot433$
    (c) $2\cdot7$, $0\cdot524$, $1\cdot91$, $0\cdot233$
11. 5; $0\cdot0232$, $0\cdot0768$, $0\cdot1232$, $0\cdot1768$
12. (b) $A = 98°\,8'$, $B = 8°\,8'$
13. $23°\,6'$, $35°\,48'$, $122°\,48'$; 376
14. $A = 44°\,44'$, $B = 104°\,16'$, $C = 31°$; $25\,900$ m^2
15. 4420, $57°\,40'$ E of N
16. $B = 77°\,30'$, $C = 60°\,10'$, $b = 452$
17. $B = 118°\,46'$, $C = 14°\,42'$, $a = 62\cdot9$, 607

4a
1. 60, 2520, 5040
2. (a) 120   (b) 360
3. 336
4. (a) 20, 10, 210   (b) 21, 15, 15
5. (a) i 56   ii 126
6. 56
7. 120
8. 3003
9. (a) 495   (b) 126
10. 6561

4b
1. $a^5 + 5a^4x + 10a^3x^2 + 10a^2x^3 + 5ax^4 + x^5$
2. $16x^4 + 32x^3y + 24x^2y^2 + 8xy^3 + y^4$
3. $27x^3 - 135x^2y + 225xy^2 - 125y^3$
4. $x^6 + 6x^4 + 15x^2 + 20 + \dfrac{15}{x^2} + \dfrac{6}{x^4} + \dfrac{1}{x^6}$
5. $243x^5 - 405x^4y + 270x^3y^2 - 90x^2y^3 + 15xy^4 - y^5$
6. $1 + 4x + 10x^2 + 16x^3 + 19x^4 + 16x^5 + 10x^6 + 4x^7 + x^8$
7. $-\dfrac{12!}{9!\,3!}2^9x^9$
8. $-\dfrac{13!}{11!}2x^2$
9. $\dfrac{8!}{4!\,4!}\times\dfrac{16}{81}a^4b^4$
10. $\dfrac{9!}{6!}\times\dfrac{125}{6x^3}$

11. $\dfrac{12!}{6!\,6!}\,a^{12}$ 12. $-\dfrac{9!}{4!\,5!}\times\dfrac{16}{x}\,;\ \dfrac{9!}{4!\,5!}\,32x$

4c

1. $1+x+x^2+x^3+\ldots$ $|x|<1$
2. $1+2x+3x^2+4x^3+\ldots$ $|x|<1$
3. $1-2x+3x^2-4x^3+\ldots$ $|x|<1$
4. $1+\dfrac{1}{2}x-\dfrac{1}{8}x^2+\dfrac{1}{16}x^3+\ldots$ $|x|<1$
5. $1+x+\dfrac{3}{2}x^2+\dfrac{5}{2}x^3+\ldots$ $|x|<\dfrac{1}{2}$
6. $\dfrac{1}{2}+\dfrac{1}{16}x+\dfrac{3}{256}x^2+\dfrac{5}{2048}x^3+\ldots$ $|x|<4$
7. $\dfrac{1}{27}-\dfrac{1}{27}x+\dfrac{2}{81}x^2-\dfrac{10}{729}x^3+\ldots$ $|x|<3$
8. $27-9x+\dfrac{1}{2}x^2+\dfrac{1}{54}x^3+\ldots$ $|x|<4\frac{1}{2}$
9. $\dfrac{1}{2}-\dfrac{1}{16}x+\dfrac{3}{256}x^2-\dfrac{5}{2048}x^3+\ldots$ $|x|<4$
10. $\dfrac{1}{3}-\dfrac{2}{9}x+\dfrac{4}{27}x^2-\dfrac{8}{81}x^3+\ldots$ $|x|<1\frac{1}{2}$
11. $\dfrac{1}{1-x}+\dfrac{2}{1+x}\,;\ 3-x+3x^2-x^3$ $|x|<1$
12. $\dfrac{3}{2+x}-\dfrac{1}{1-2x}\,;\ \dfrac{1}{2}-2\frac{3}{4}x-3\frac{5}{8}x^2-8\frac{3}{16}x^3$ $|x|<\dfrac{1}{2}$
13. $\dfrac{1}{3(1+x)}+\dfrac{8}{3(1-2x)}-\dfrac{1}{(1-2x)^2}\,;\ 2+x-x^2-11x^3$ $|x|<\dfrac{1}{2}$
14. $\dfrac{1}{1-x}+\dfrac{2}{(1-x)^2}-\dfrac{1}{2-x}\,;\ 2\frac{1}{2}+4\frac{3}{4}x+6\frac{7}{8}x^2+8\frac{15}{16}x^3$ $|x|<1$

4d

6. $9\cdot8995$ 7. $10\cdot0100$ 8. $9\cdot9933$
9. $7\cdot9896$ 10. $1\cdot0934$ 12. -1% error
13. $3\cdot107$ 14. $7\cdot5\%$ increase   15. $0\cdot202$
16. $4\cdot123$ 17. $6\frac{1}{2}\%$ too small
18. $1-x-\dfrac{x^2}{2}\,;\ 1-3x+6x^2\,;\ 1-4x+8\frac{1}{2}x^2$ $|x|<\frac{1}{2}$
19. 7% decrease   20. (a) $4-\dfrac{43}{30}x$ (b) $1+\dfrac{65}{108}x$

4f 1. $1 + \dfrac{x}{2} + \dfrac{x^2}{8} + \dfrac{x^3}{48} + \ldots$ All values of x

2. $1 + x + \dfrac{3}{2}x^2 + \dfrac{7}{6}x^3 + \ldots$ All values of x

3. $1 - 3x + \dfrac{7}{2}x^2 - \dfrac{13}{6}x^3 + \ldots$ All values of x

4. $1 + 2\frac{1}{2}x + 3\frac{3}{8}x^2 + 3\frac{19}{48}x^3$ $|x| < 1$

5. $2x - 2x^2 + \dfrac{8}{3}x^3 - 4x^4$ $|x| < \dfrac{1}{2}$

6. $2x - x^2 + \dfrac{2}{3}x^3 - \dfrac{1}{2}x^4$ $|x| < 1$

7. $\log 2 + x - \dfrac{x^2}{2} + \dfrac{x^3}{3}$ $|x| < 1$

8. $5x - \dfrac{13}{2}x^2 + \dfrac{35}{3}x^3 - \dfrac{97}{4}x^4$ $|x| < \dfrac{1}{3}$

9. $x + \dfrac{5}{2}x^2 + \dfrac{19}{3}x^3 + \dfrac{65}{4}x^4$ $|x| < \dfrac{1}{3}$

10. $1 + 4x + \dfrac{9}{2}x^2 + \dfrac{8}{3}x^3$ All values of x

11. (a) i $6\cdot6285$ ii $\bar{6}\cdot7341$
    (b) i $3\cdot808$ ii 2057 iii $0\cdot001\,798$
    (c) $80\cdot22$

12. (a) $1 + \dfrac{1}{3!} + \dfrac{1}{5!} + \dfrac{1}{7!} + \ldots$

    (b) $2\left[1 + \dfrac{1}{2!}\left(\dfrac{1}{2}\right)^2 + \dfrac{1}{4!}\left(\dfrac{1}{2}\right)^4 + \dfrac{1}{6!}\left(\dfrac{1}{2}\right)^6 + \ldots \right]$

13. (a) $1\cdot005$ (b) $0\cdot100$ 14. $1\cdot098\,612$

15. $0\cdot693, -1\cdot099$ 16. $-0\cdot406, -0\cdot693$

17. $2, -5$ 18. $2 + 9x^2 + \dfrac{81}{12}x^4 + \ldots; \pm0\cdot1$

20. $1 + 2x + 2x^2; 1 + 3x + 6x^2; 1\cdot051$

Miscellaneous exercises 4

1. (a) $1 + x + \dfrac{3}{2}x^2 + \dfrac{5}{2}x^3 \ldots$ $|x| < \dfrac{1}{2}$

    (b) $1 - \dfrac{3}{2}x$

2. (a) $2+\dfrac{x}{4}-\dfrac{x^2}{32}+\dfrac{5}{768}x^3\ldots$ $|x|<\dfrac{8}{3}$

  (b) $1\cdot065$

3. (a) $5x^4,\ \dfrac{1}{4}x^{-\frac{3}{4}},\ -\dfrac{2}{x^3}$ (b) 1

4. $\dfrac{2}{1+x}+\dfrac{1}{1+2x}$; $3-4x+6x^2-10x^3\ldots$ $|x|<\dfrac{1}{2}$

5. (a) i $1+\dfrac{x}{2}-\dfrac{x^2}{8}+\dfrac{x^3}{16}\ldots$ ii $1-\dfrac{x}{2}+\dfrac{3}{8}x^2-\dfrac{5}{16}x^3+\ldots$ 1

  (b) $0\cdot50\%$

6. (a) $1-4x+6x^2-4x^3$; $256+1024x+1792x^2+1792x^3$
     $A=256,\ B=0,\ C=-1368,\ D=-246$

  (b) $9\cdot798$

7. (a) i 161 ii 84 (b) 210

8. (a) $|x|<1$; $n+1$; $\dfrac{32}{25}$ (b) $1\cdot0139$

9. (a) $0\cdot833$ (b) $0\cdot936$ (c) $0\cdot9$ (d) $1\cdot58$

10 (a) $91\cdot8$ (b) $0\cdot0229$ s

11. (a) $3\cdot606$ (b) $4\cdot6\%$ decrease

12. (a) $1-\tfrac{1}{2}x^2-\tfrac{1}{8}x^4,\ 1-\tfrac{1}{2}x^2+\tfrac{3}{8}x^4$ $|x|<1$; $0\cdot99005$

  (b) P decreases by $2\cdot6\%$

13. $0\cdot6931$

14. $|x|<1,\ -\infty<y<1$

15. 2% decrease

16. (a) $-6-6c+3c^2$ (b) $1\cdot010$

18. $x-\dfrac{x^2}{2}+\dfrac{x^3}{3}-\ldots$ $|x|<1$; $t=20\left(x+\dfrac{3}{2}x^2+\dfrac{7}{3}x^3+\ldots\right)$

19. (a) $3,9$ (b) $d=1$ 20. $0\cdot4055$

21. (a) $1\cdot0182$ (b) $-0\cdot865$ 22. $1\cdot89$

5a 1. $-1,-1,1,\ x^2-x-1,\ x^2+(2h-3)x+h^2-3h+1$

  2. 4, 2, indeterminate   3. 3

  4. 7   5. 6   6. 3

  7. $\tfrac{3}{5}$ 8. $\tfrac{1}{2}$ 9. $2x$

  10. $6x$ 11. $2x-2$ 12. $3x^2+2$

5c 1. 5   2. $4x-3$ 3. $2x$

  4. $2\cos 2x$ 5. $-\dfrac{2}{x^{-3}}$ 6. $-\dfrac{2}{(2x+1)^2}$

7. $7x^{2.5}$

8. $\dfrac{1.5}{S^{2.5}} - \dfrac{2.6}{S^{3.6}}$

9. $\dfrac{2}{3\sqrt[3]{t}}$

10. $\cos\theta - \sin\theta$

11. $7v^{0.4}$

12. $-\dfrac{1.2}{p^{2.2}}$

13. $2\cos t + 3\sin t$

14. $-\dfrac{5}{2\sqrt{y^7}}$

15. $2 - \cos x$

16. $9z^2 + 2\sin z$

17. $2x - 4\cos x + 2\sin x$

18. $-2(\sin\theta + \cos\theta)$

19. $3\cos v - \dfrac{10}{v^2}$

20. $5.6t^{0.4} + \dfrac{\sin t}{3}$

21. (a) $3x^2 - 4x + 1$ (b) $2\cos\theta + \sin\theta$

22. $2x - 2$

5e

1. $5\cos x + 3\sin x$

2. $3 - \dfrac{8}{t^3}$

3. $x^2 - \dfrac{5}{2} + \dfrac{3}{2x^2} - \dfrac{4}{x^3}$

4. $6t + \dfrac{4}{t^3}$

5. $\dfrac{1}{2\sqrt{x}} - \dfrac{1}{2\sqrt{x^3}}$

6. $1 + 6\sqrt{x} + 8x$

7. $1 + \dfrac{2}{x^3}$

8. $3x^2 + 3 - \dfrac{3}{x^2} - \dfrac{3}{x^4}$

9. $-\dfrac{1}{x^2} - 1 + \sin x$

10. $3x^2 - 2x + 2$

11. (a) $6x - \dfrac{2}{x^3}$ (b) $12x + 2$

14. $x + y = 2$

15. $9x + y = 14$

16. $(\tfrac{1}{3}, 4\tfrac{13}{27})$ $(-1, 3)$

5g

1. $\cos x - x\sin x$

2. $2\sin x + (2x - 1)\cos x$

3. $(8x^3 - 3)(x^3 + 2x - 1) + (2x^4 - 3x)(3x^2 + 2)$

4. $\dfrac{1}{2\sqrt{x}}(x^2 - x - 1) + (\sqrt{x} + 1)(2x - 1)$

5. $3x^2\sin x + x^3\cos x$

6. $\tan x + (x + 2)\sec^2 x$

7. $\dfrac{x\cos x - \sin x}{x^2}$

8. $\dfrac{3 - 3x^2 + 2x}{(x^2 + 1)^2}$

9. $-\dfrac{3}{(x - 1)^2}$

10. $\dfrac{2x\cos x + (x^2 - 1)\sin x}{\cos^2 x}$

11. $\dfrac{(x+1)\sec^2 x-\tan x}{(x+1)^2}$

12. $\dfrac{2x^2-18x+3}{(2x^2-3)^2}$

13. $\cos t-t\sin t$

14. $(2t+2)\sin t+(t^2+2t-1)\cos t$

15. $\dfrac{(x^2-1)\,3\cos x-6x\sin x}{(x^2-1)^2}$

16. $\cos^2\theta-\sin^2\theta=\cos 2\theta$

17. $2x\tan x+x^2\sec^2 x$

18. $\dfrac{(3x^2-1)\sin x-(x^3-x)\cos x}{\sin^2 x}$

19. $\dfrac{2x^3+9x^2+12x+9}{(x+1)^2}$

20. $\dfrac{(\theta-1)\sec^2\theta-\tan\theta}{(\theta-1)^2}$

21. $-\dfrac{1}{\sqrt{y}(\sqrt{y}-1)^2}$

22. $-\dfrac{(2z+1)}{(z^2+z)^2}$

23. $(x^4+1)\sec^2 x+4x^3\tan x$

24. $x^3\sin x+3x^2(1-\cos x)$

25. $2x\sec x+(x^2-1)\sec x\tan x$

26. $2x\cot x-x^2\operatorname{cosec}^2 x$

27. $\dfrac{x^2\cos x-x\cos x-\sin x}{(x-1)^2}$

28. $\dfrac{(x-x^2)\left[\tan x+(x+1)\sec^2 x\right]-(1-2x)(x+1)\tan x}{(x-x^2)^2}$

29. $-x\cos x-2\sin x$

30. $4x\cos x+2\sin x-x^2\sin x$

31. $\dfrac{6\sin x-2x\cos x-x^2\sin x-2x\cos x}{x^4}$

32. $\dfrac{-x^2\cos x+2\cos x+2x\sin x}{x^3}$

5h

1. $4(x^2+x)^3(2x+1)$

2. $-24x(3-x^2)^5$

3. $6x\cos(3x^2+1)$

4. $4\sec^2 4x$

5. $-4x\sin(2x^2-1)$

6. $11(x^2+x)^{10}(2x+1)$

7. $-16x(1-x^2)^7$

8. $3(3x^2-4x+5)^2(6x-4)$

9. $\dfrac{3x-2}{\sqrt{(3x^2-4x+5)}}$

10. $\dfrac{1-2x}{(x^2-x-1)^2}$

11. $3\cos(3x+2)$

12. $2\sec^2 2x$

13. $3\sec 3x\tan 3x$

14. $6(x^2+x)^5(2x+1)$

15. $12(x^2+x)^{11}(2x+1)$

16. $7a(ax+b)^6$

17. $4\sec^2(4x+3)$

18. $21(3x+1)^6$

19. $-\dfrac{21}{(3x+1)^8}$

20. $\dfrac{3}{2\sqrt{(3x+1)}}$

21. $10\sin^9 x\cos x$

22. $-11\cos^{10} x\sin x$

23. $10x(x^2+1)^4$

24. $5\cos x(\sin x+1)^4$

25. $-\dfrac{2\cos x}{(1+\sin x)^3}$ 26. $3\sin^2 x \cos x$

27. $-3\cos^2 x \sin x$ 28. $-3\sin 3x$

29. $-4\,\mathrm{cosec}\,4x\cot 4x$ 30. $-4\cot^3 x\,\mathrm{cosec}^2 x$

31. $\dfrac{x^2}{(x^3+8)^{\frac{2}{3}}}$ 32. $4\cos 4x$

33. $-2x\sin(x^2-1)$ 34. $\dfrac{1}{2\sqrt{(x+1)}}$

35. $3\sec^2 3x$ 36. $8\sec 2x\tan 2x$

37. $-\dfrac{1}{2(x+2)^{\frac{3}{2}}}$ 38. $-6\,\mathrm{cosec}^2 3x$

39. $4\sec^2(4x+3)$ 40. $(x-1)^9[10\sin x+(x-1)\cos x]$

41. $2x\cos 3x+3\sin 3x-3x^2\sin 3x$

42. $\dfrac{\sin 4x-4x\cos 4x}{\sin^2 4x}$ 43. $-\dfrac{3\cos 3x}{\sin^2 3x}$

44. $\dfrac{x\cos(2x+1)-2(x^2+4)\sin(2x+1)}{\sqrt{(x^2+4)}}$

45. $\dfrac{\sec 3x\,[3(2x-3)\tan 3x-1]}{(2x-3)^{\frac{3}{2}}}$

5j

1. $2,\ 2t,\ \dfrac{1}{t}$ 2. $-2t,\ 2t,\ -1$

3. $2t,\ 6t^2,\ \dfrac{1}{3t}$ 4. $-\dfrac{c}{t^2},\ c,\ -\dfrac{1}{t^2}$

5. $2a,\ 2at,\ \dfrac{1}{t}$ 6. $1-\dfrac{1}{t^2},\ 2t-\dfrac{2}{t^3},\ \dfrac{t}{2(t^2+1)}$

7. $b\cos t,\ -a\sin t,\ -\dfrac{b}{a}\cot t$ 8. $\sec^2 t,\ \sec t\tan t,\ \mathrm{cosec}\,t$

9. $3a\sin^2 t\cos t,\ -3a\cos^2 t\sin t,\ -\tan t$

10. $\dfrac{1}{3y^2-1}$ 11. $\dfrac{1}{3(y+1)^2}$

12. $\sec y$ 13. $\dfrac{2}{y+1}$

14. (a) $-\dfrac{x}{y}$ (b) $\dfrac{x+1}{3-y}$ (c) $-\dfrac{3x^2+2y}{2x+3y^2}$

5k

1. $3e^{3x}$ 2. e^x-e^{-x}

3. $2xe^{x^2}$ 4. $\dfrac{2}{2x+3}$

5. $-\tan x$

6. $\dfrac{4x}{2x^2-1}$

7. $-\dfrac{1}{1-x}$

8. $e^x(\sin x+\cos x)$

9. $4^x \log_e 4$

10. $-\dfrac{2}{e^{2x-1}}$

11. $e^x(x+1)$

12. $e^{2x}(2\cos 3x-3\sin 3x)$

13. $x(1+2\log x)$

14. $e^{2x}\left(2\log\tan x+\dfrac{2}{\sin 2x}\right)$

15. $-\dfrac{1}{2-x}-\dfrac{1}{2+x}$

16. $\dfrac{e^{ax}(a\sin bx-b\cos bx)}{\sin^2 bx}$

17. $\dfrac{2x(1-x)}{e^{2x}}$

18. $\dfrac{1-2x\log x}{xe^{2x}}$

19. $e^{ax}x^{n-1}(n+ax)$

20. $e^{\tan x}(1+x\sec^2 x)$

21. $(\cos x-\sin x)\,e^{\cos x+\sin x}$

22. $\dfrac{\sin x+\cos x}{\sin x-\cos x}$

5l 1. $\dfrac{(x+2)(x-4)}{(x-1)(x-2)}\left(\dfrac{1}{x+2}+\dfrac{1}{x-4}-\dfrac{1}{x-1}-\dfrac{1}{x-2}\right)$

2. $\dfrac{(x-1)^3}{(x+1)^2(x-2)}\left(\dfrac{3}{x-1}-\dfrac{2}{x+1}-\dfrac{1}{x-2}\right)$

3. $\dfrac{(x-1)^{\frac{1}{2}}}{(x+1)^{\frac{1}{3}}(x+2)^{\frac{1}{3}}}\left\{\dfrac{1}{2(x-1)}-\dfrac{1}{3(x+1)}-\dfrac{1}{3(x+2)}\right\}$

4. $(x+1)^x\left\{\dfrac{x}{x+1}+\log(x+1)\right\}$

5. $\sqrt[x]{(1+x)}\left\{\dfrac{1}{x(x+1)}-\dfrac{\log(1+x)}{x^2}\right\}$

6. $x^{2x-2}(2x\log x+2x-1)$

Miscellaneous exercises 5

1. (a) $\dfrac{x^2-2x-1}{(x-1)^2}$ (b) $\dfrac{4x}{1+2x^2}$

    (c) $2\sin 2x\cos x+4\sin x\cos 2x$ (d) $2xe^{\frac{1}{2}x^2}$

2. (a) i $-\dfrac{(x^2+1)}{(x^2-1)^2}$ ii $e^x(\sin x+\cos x)$ iii $\dfrac{2x}{x^2-1}$

    (b) 8, $-\frac{1}{8}$

3. (a) i $\dfrac{2x}{\sqrt{(2x^2-3)}}$ ii $-\dfrac{1}{(2x-3)^2}$ iii $15\sin^2 5x\cos 5x$

4. (a) $2\cos 2\theta$ (b) i $\dfrac{1}{(4-2x)^{\frac{3}{2}}}$ ii $e^{-3x}(\cos x-3\sin x)$

   iii $\dfrac{3\cos 2x}{\cos^4 x}$ iv $(x-1)+2x\log(2x+2)$

5. (b) i $\dfrac{1+4x}{(1-2x^2)^2}$ ii $3\cos 6x$ iii $\dfrac{1}{2x-5}$ iv $-e^{-ax}(a\cos x+\sin x)$

6. (a) i $12x^2(x^3+2)^3$ ii $2\sin(1-2x)$ iii $\dfrac{12xe^{2x}}{(6x+3)^2}$

7. (a) $-\dfrac{1+3x}{2\sqrt{(1+x)}}$ (b) $\dfrac{2x-2-2x^2}{e^{2x}}$ (c) $\dfrac{2}{2x-1}$

8. (a) $\dfrac{4}{(1-x)^3}$

9. (a) i $2xe^{2x}(1+x)$ ii $5x(x^2+a^2)^{\frac{3}{2}}$ iii $-a\tan ax$

10. (b) $\dfrac{(x+1)^3(x-2)^2}{(x+2)^2\cdot\sqrt{(x+4)}}\left(\dfrac{3}{x+1}+\dfrac{2}{x-2}-\dfrac{2}{x+2}-\dfrac{1}{2(x+4)}\right)$ (c) $9y=2x+5$

11. (a) $2\sin x\cos 2x+\sin 2x\cos x$
    $4\cos x\cos 2x-5\sin x\sin 2x$
    (b) $3x^2\sin x+x^3\cos x$
    $(6x-x^3)\sin x+6x^2\cos x$

12. (a) 6   (b) $\dfrac{(x+1)^3(x-2)}{(x-3)^2(x+4)}\left(\dfrac{3}{x+1}+\dfrac{1}{x-2}-\dfrac{2}{x-3}-\dfrac{1}{x+4}\right)$ (c) $3y=2x-1$

13. (a) i $-\dfrac{6x}{(x^2-4)^2}$ ii $x+2x\log x$ iii $6\tan(3x-5)\sec^2(3x-5)$

14. (a) i $-\dfrac{\cos x}{\sin^2 x}$ ii $e^{2x}(3\cos 3x+2\sin 3x)$ iii $5\tan\left(5x+\dfrac{\pi}{4}\right)$

15. (a) i $\dfrac{3}{x}$ ii $\dfrac{\sin 2x}{(1+\sin x)^2}$ iii $e^{\frac{1}{2}x}(\frac{1}{2}\tan x+\sec^2 x)$
    (b) $y=2\log x+x-1$

16. (a) i $\dfrac{(1+x^2)}{2\sqrt{x}(1-x^2)^{\frac{3}{2}}}$ ii $e^{-x}(2\cos 2x-\sin 2x)$ iii $\dfrac{4}{\cos 4x}$

17. (a) i $-\dfrac{x}{\sqrt{(1-x^2)}}$ ii $\dfrac{2x^2-6x+2}{(1-x^2)^2}$ iii $2\operatorname{cosec} 2x$

    (c) $\dfrac{4}{5}$

18. (a) i $-\dfrac{41}{(5x-4)^2}$ ii $\dfrac{1}{2}\cot\dfrac{x}{2}$ iii $e^x(\cos 2x-2\sin 2x)$

19. (a) i $\dfrac{1}{(3-2x)^2}$ ii $-\dfrac{4x}{(1+x^2)^2}$ iii $-x^2 e^{-x}$

20. (a) i $-\dfrac{1}{x}$ ii $-\dfrac{1}{2}\cos\left(\dfrac{\pi}{6}-\dfrac{x}{2}\right)$ iii $\dfrac{2x}{(x^2+1)^2}$

6c

1. (a) 10 m s^{-1}, -6 m s^{-2} (b) 3.49 s   (c) 38.4 m
2. 16 rad s^{-1}, 14 rad s^{-1}, 16 s, 20.4
3. (a) 12.57 m^2 s^{-1} (b) 18.85 m^2 s^{-1}
4. 0.252
5. 0.309 cm s^{-1}
6. 15 cm s^{-1}
7. (a) 0.088 dm s^{-1} (b) 0.032 dm s^{-1}
8. (a) Min. $(\frac{1}{3}, \frac{22}{27})$, max. $(-1, 2)$ (b) 2.45, 3.46
9. (a) Max. $y = 5$, min. $y = 1$ (b) $\sqrt[4]{\dfrac{4Tg}{\rho}}$
10. (a) Max. -3, min. -2 (b) Max. $\left(14° 29', \dfrac{9}{8}\right)$, min. $\left(\dfrac{\pi}{2}, 0\right)$
12. Max. 10 at $71° 34'$, min. 2 at $161° 34'$
13. $4\frac{1}{2}$ by $13\frac{1}{2}$
14. (a) Max. $(3, 12\log 3 - 3)$ (b) Max. $\left(30°, \dfrac{3\sqrt{3}}{4}\right)$, min. $\left(150°, -\dfrac{3\sqrt{3}}{4}\right)$
15. 6.34 cm, 6.34 cm
16. 0.168 m^2
17. (a) -0.0011 (b) 0.127
18. $\dfrac{\delta x}{2\sqrt{x}}$; $50\dfrac{\delta x}{x}$ %; $-\dfrac{1}{9}$ %
19. 0.131
20. 1.78 mm away

6f

1. (a) $y = 2x+3$ (b) $3y = -x-3$ (c) $y = 3x+2$ (d) $2y = x-5$
   (e) $2x+y = 1$
2. $x+y = 2$, $y = x$ 3. 1; $x+y+2 = 0$
4. $\dfrac{1}{t}$; $ty = x+2t^2$, $tx+y = 4t+2t^3$
5. $2x+3y = 12$ 6. $9x+y = 6$, $9y = x+26\frac{2}{3}$
7. 2.25 8. 2.37
9. -1.33 10. 0.567
11. (a) $2x - \dfrac{2^3 x^3}{3!} + \dfrac{2^5 x^5}{5!} - \dfrac{2^7 x^7}{7!} + \dots$

(b) $1 - \dfrac{a^2x^2}{2!} + \dfrac{a^4x^4}{4!} - \dfrac{a^6x^6}{6!} + \ldots$

(c) $1 - \dfrac{x^2}{2^2 2!} + \dfrac{x^4}{2^4 4!} - \dfrac{x^6}{2^6 6!} + \ldots$

12. $x + x^2 + \dfrac{x^3}{3} + \ldots$

13. $x - \dfrac{x^3}{3!} + \dfrac{x^5}{5!} - \dfrac{x^7}{7!}$; 0·099 833

14. $x - \dfrac{x^2}{2} - \dfrac{x^3}{6}$

6h

1. y, x

2. $2xy^3 + 2, 3x^2y^2 + 1$

3. $\dfrac{3}{y}, -\dfrac{3x}{y^2}$

4. $3x^2y - \dfrac{2y}{x^3}, x^3 + \dfrac{1}{x^2} + \dfrac{1}{y^2}$

5. $3\cos(3x + 2y), 2\cos(3x + 2y)$

6. $2\cos 2x \cos 3y, -3\sin 2x \sin 3y$

7. $-\dfrac{RT}{v^2}, \dfrac{R}{v}$

8. $\dfrac{r}{t}, \dfrac{P}{t}, -\dfrac{Pr}{t^2}$

9. $\dfrac{\pi}{\sqrt{(lg)}}, -\pi\sqrt{\dfrac{l}{g^3}}$

10. $\dfrac{fQ^2}{10d^5}, \dfrac{2flQ}{10d^5}, -\dfrac{flQ^2}{2d^6}$

13. $-7\cdot2$ cm^3

14. -90 cm^3

15. $-0\cdot39$ amp

16. $\delta Q = 1\cdot82H^{\frac{3}{2}}\delta L + 2\cdot7LH^{\frac{1}{2}}\delta H$; $-0\cdot044$ m^3 s^{-1}

Miscellaneous exercises 6

1. (a) $\dfrac{1}{2\pi}$ (b) $\dfrac{1}{2}$

2. (a) Max. 1, min. 25

3. (a) 30 m s^{-1}, 8 m s^{-2} (b) $2\frac{1}{6}$ m s^{-1}

4. (a) Max.-1, min.$-\frac{1}{4}$ (b) $0\cdot378$ mm s^{-1}

5. $(\frac{1}{2}, 1\frac{11}{16})$ min.; $(0, 1\frac{5}{6})$ max.; $(-1\frac{1}{2}, 0\cdot98)$ min.

6. (a) $\dfrac{(a+b)^2}{16w^2} + c^2at - \dfrac{a+b}{4w^2}$; $230\frac{1}{3}$, $x = \dfrac{4}{3}$ (b) $+2$

7. (a) $30°$ (b) $\dfrac{1}{\sqrt{2}}$

8. $x = 16, y = 20$

9. (a) $a = 3, b = -12$; max. 23, min. -4; $(-\frac{1}{2}, 9\frac{1}{2})$

(b) $-\dfrac{1}{30}$ amperes per minute

10. (a) $\pm3\cdot33$ (b) £140 250; £38 750, 50

11. $y - y_1 = m(x - x_1)$, $x + 2y = 5$

12. (a) $0\cdot0032$ (b) -216

13. (a) $-2\cos(2x-3y)$, $3\cos(2x-3y)$ (b) $\dfrac{200}{t}$, $-\dfrac{200r}{t^2}$

14. 0·35

15. 1·65

16. −0·57

17. 2·09

18. 48 MN m^{-2}; 7·64 MN m^{-2}

19. $\delta q \simeq \dfrac{16w}{\pi}\left(\dfrac{\delta R}{d^3} - \dfrac{3R}{d^4}\delta d\right)$; -3%

20. 7% decrease

7a

1. $n\pi + (-1)^n \dfrac{\pi}{4}$; $45°$, $135°$

2. $n\pi \pm \dfrac{\pi}{6}$; $30°$, $150°$, $210°$, $330°$

3. $n\pi + \dfrac{\pi}{3}$; $60°$, $240°$

4. $2n\pi \pm \dfrac{2}{3}\pi$; $120°$, $-120°$

5. $n\pi + \dfrac{3}{4}\pi$; $135°$, $315°$

6. $n\pi + \dfrac{\pi}{4}$; $45°$, $225°$

7. $2n\pi \pm \dfrac{\pi}{3}$; $60°$, $300°$

8. $n\pi + (-1)^n \dfrac{\pi}{3}$; $60°$, $120°$

9. (a) $(2n+1)\pi + \dfrac{\pi}{4}$; $225°$ (b) $2n\pi + \dfrac{7\pi}{6}$, $210°$

10. $n\pi + (-1)^n \dfrac{\pi}{6}$; $30°$, $150°$

11. $2n\pi \pm \dfrac{2}{3}\pi$

12. $n\pi + (-1)^n \dfrac{\pi}{3}$

13. $360n° \pm 51° \, 50'$

14. $n\pi + (-1)^n \dfrac{\pi}{10}$ or $n\pi - (-1)^n \dfrac{3\pi}{10}$

15. $n\pi + \dfrac{\pi}{4}$ or $n\pi + \dfrac{\pi}{3}$

16. $n\dfrac{\pi}{5}+(-1)^n\dfrac{\pi}{20}$ 17. $\dfrac{2}{3}n\pi\pm\dfrac{\pi}{6}$

18. $n\dfrac{\pi}{4}$ or $\dfrac{(2n+1)\pi}{10}$ 19. $2n\pi$ or $\dfrac{(2n+1)\pi}{5}$

20. $2n\pi$ or $\dfrac{2n\pi}{9}$ 21. $\left(n+\dfrac{1}{2}\right)\dfrac{\pi}{4}$

22. $\dfrac{n\pi}{4}$ or $\dfrac{1}{3}\left(2n\pi\pm\dfrac{\pi}{3}\right)$ 23. $\left(n+\dfrac{1}{2}\right)\dfrac{\pi}{4}$ or $\left(2n\pm\dfrac{1}{3}\right)\dfrac{\pi}{3}$

24. $\left(n+\dfrac{1}{2}\right)\dfrac{\pi}{2}$ or $2n\pi$ 25. $n\dfrac{\pi}{3}$ or $\left(n\pm\dfrac{1}{3}\right)\pi$

7b 1. $(360n-90)^\circ$ or $360n^\circ+46^\circ\,24'$; $46^\circ\,24'$, 270°
2. $180n^\circ+(-1)^n\,68^\circ\,12'-21^\circ\,48'$; $46^\circ\,24'$, 90°
3. $n\pi+(-1)^n\dfrac{\pi}{4}-\dfrac{\pi}{3}$; 75°, 345°
4. $2n\pi,\ 2n\pi-\dfrac{2}{3}\pi$; 0°, 240°, 360°
5. $360n^\circ\pm73^\circ\,54'+56^\circ\,19'$; $130^\circ\,13'$, $342^\circ\,25'$
6. $180n^\circ+(-1)^n\,36^\circ\,2'-28^\circ\,4'$; $7^\circ\,58'$, $115^\circ\,54'$
7. $360n^\circ+125^\circ\,40'$ or $360n^\circ-19^\circ\,24'$; $125^\circ\,40'$, $340^\circ\,36'$
8. $360n^\circ+62^\circ\,29'$, $360n^\circ+86^\circ\,21'$; $62^\circ\,29'$, $86^\circ\,21'$
9. $180n^\circ+61^\circ\,56'+(-1)^n\,14^\circ\,28'$; $76^\circ\,24'$, $227^\circ\,28'$
10. $2n\pi$ or $2n\pi+\dfrac{\pi}{2}$; 0°, 90°, 360°

7c 1. $\dfrac{\pi}{3}, -\dfrac{\pi}{6}, \dfrac{\pi}{4}, \dfrac{\pi}{6}$, $19^\circ\,28'$, $48^\circ\,12'$

3. $100^\circ\,32'$ 5. $\dfrac{\pi}{2}$

Miscellaneous exercises 7

1. (a) $76^\circ\,24'$, $227^\circ\,28'$ (b) 210°, 270°, 330°
   (c) 36°, 108°, 180°, 252°, 324°, 120°
2. (a) $14^\circ\,28'$, $165^\circ\,32'$ (b) $53^\circ\,8'$, $292^\circ\,42'$
3. (a) $R = 17$, $\alpha = 60^\circ$ (b) 150°, 270°
4. (b) $A = 52^\circ\,47'$, $B = 7^\circ\,13'$ (c) 120°, 240°
5. $x = 25\sin(t-0.284)$ (a) 0.928 s (b) 4.07 s
6. (a) $18^\circ\,26'$, $71^\circ\,34'$

7. $x = 45°$ and $135°$

8. $0.37 \sin(\theta - 71° 4')$; $+0.37$ at $161° 4'$ and -0.37 at $341° 4'$; $13\pi + 1.24$ rad

9. (b) $84° 16'$, $123° 35'$, $246° 25'$, $275° 44'$

10. (b) $25 \sin(x + 16° 16')$; $200° 36'$ and $306° 52'$

11. (a) $2\cos^2 \theta - 1$, $4\cos^3 \theta - 3\cos\theta$ (b) 1.979, -0.347, -1.532

12. $2.65 \cos(\theta + 40° 54')$; 2.65, $319° 6'$, $26° 55'$, $251° 7'$

13. (a) $30°$, $52\frac{1}{2}°$, $142\frac{1}{2}°$ (b) $0°$, $45°$, $105°$, $165°$, $180°$

14. $2\cos(\theta - 60°)$; $6° 52'$

15. $1.3 \sin(\theta - 22° 37')$ (a) $52° 37'$, $172° 32'$

16. $5\cos(x + 0.927)$ (a) -0.927, 5.36 (b) 2.22 (c) 0.644, 3.79
    (d) 0.120, 4.31

17. (a) 0.343, -1.094; $34° 27'$ (b) $x = 0.644$, $y = 0.404$ rad

18. (a) 4 (b) $x = n\pi \pm \dfrac{\pi}{6}$; $30°$, $150°$, $210°$, $330°$

    (c) $25 \sin(\theta + 73° 44')$; max. 25 at $\theta = 16° 16'$

19. (b) $26.4 \sin(\theta - 52° 42')$; $142° 42' + 360n°$

20. (a) $\dfrac{\sqrt{3}}{2} \sin x + \dfrac{5}{2} \cos x = \sqrt{7} \sin(x + 70° 52')$

8a

1. $\dfrac{x^3}{3} + 3x + \dfrac{1}{x} + c$

2. $\dfrac{2}{3} x^{\frac{3}{2}} + 2\sqrt{x} + c$

3. $\dfrac{x^4}{4} - \dfrac{1}{2x^2} + c$

4. $\dfrac{x^3}{3} + x^2 + x + c$

5. $\dfrac{4x^3}{3} - 6\sqrt{x} + c$

6. $x + \dfrac{4}{3} x^{\frac{3}{2}} + \dfrac{x^2}{2} + c$

7. $2x^{\frac{3}{2}} - \dfrac{2}{3} x^{\frac{3}{2}} + c$

8. $\dfrac{e^{2x}}{2} - \log x + c$

9. $4\log x + \dfrac{1}{x} + c$

10. $\dfrac{e^{3x}}{6} + \dfrac{e^{-3x}}{6} + c$

11. $\dfrac{\sin 4x}{4} + c$

12. $\tan x + \cos x + c$

13. $5e^{0.2x} + \dfrac{x^2}{2} + c$

14. $\dfrac{\sin 3x}{3} + \dfrac{\cos 2x}{2} + c$

15. $-2\cos \dfrac{x}{2} + c$

16. $e^x + 2\log x + \dfrac{3}{e^x} + c$

17. $4\tan x + \cot x + c$

18. $\frac{1}{2}\log x + c$

19. $\dfrac{\sin bx}{b} - \dfrac{\cos ax}{a} + c$

20. $\dfrac{x^3}{3} + 2x - \dfrac{1}{x} + c$

21. $\dfrac{2}{3}t^{\frac{3}{2}}+2\sqrt{t}+c$

22. $\dfrac{\sin 2\theta}{2}-\dfrac{\cos 3\theta}{3}+c$

23. $\tan\theta+\dfrac{\sin 3\theta}{3}+c$

24. $-\dfrac{5}{2v^{0\cdot4}}-\log v+c$

25. $e^{t}+\dfrac{1}{2e^{2t}}+\log t+c$

26. $-\dfrac{\cos 100\pi t}{100\pi}-\dfrac{\sin 50\pi t}{50\pi}+c$

27. $\dfrac{2}{3}r^{\frac{3}{2}}+\dfrac{2}{5}r^{\frac{5}{2}}+c$

28. $3t+t^{2}-\dfrac{t^{3}}{3}+c$

29. $4t+\dfrac{8}{3}t^{\frac{3}{2}}+\dfrac{t^{2}}{2}+c$

30. $\cos x-\cot x+c$

8c
1. $2\frac{2}{3}$	2. 21	3. $\frac{1}{2}$	4. 16
5. $6\frac{5}{6}$	6. 4	7. 3·06	8. 4
9. 0·37	10. 1·12	11. 0·11	12. 1·98
13. 0·406	14. 1·50	15. 0·423	16. $-\frac{1}{3}$
17. 1	18. $3\frac{1}{3}$	19. 0·654	20. $\dfrac{2}{\omega}$
21. 65·8	22. $2\frac{5}{6}$	23. 9·10	24. 0·144
25. 0·413	26. 0·0245	27. 1·02	28. 2·08
29. $-6\frac{5}{6}$	30. 23·95	31. $-\frac{4}{3}$	32. $20\frac{2}{3}$
33. 0·293	34. 0	35. $9\frac{1}{3}$	36. 3·63
37. 2·77	38. 0·667	39. 0·491	40. 8

8d The constant of integration is omitted.

1. $\dfrac{(x+4)^{3}}{3}$

2. $\dfrac{(x+1)^{11}}{11}$

3. $-\dfrac{(2-x)^{4}}{4}$

4. $\dfrac{(2x+3)^{\frac{3}{2}}}{3}$

5. $-\dfrac{1}{4(2x+1)^{2}}$

6. $\frac{1}{2}\log(2x+1)$

7. $\dfrac{e^{2x}}{2}$

8. $-\dfrac{e^{-3x}}{3}$

9. $-\dfrac{\cos 3x}{3}$

10. $\dfrac{\sin 4x}{4}$

11. $\dfrac{(x^{2}+1)^{6}}{6}$

12. $\dfrac{(x^{2}+1)^{4}}{8}$

13. $\dfrac{2}{3}(x^{2}+x)^{\frac{3}{2}}$

14. $\dfrac{(ax^{2}+bx)^{5}}{5}$

15. $\dfrac{\sin(3x+2)}{3}$

16. $\sin(x^{2}+x)$

17. $\cos(4-x)$

18. $-\dfrac{1}{6(x^{6}+1)}$

19. $e^{5x^{2}-x}$

20. $-\frac{1}{2}\sqrt{(a^{4}-x^{4})}$

21. $-\dfrac{\cos^{4}x}{4}$

22. $\dfrac{\sin^3 x}{3}$ 23. $\dfrac{\tan^2 x}{2}$ 24. $\frac{1}{2}(\log x)^2$

25. $-\dfrac{\sin (2-3x)}{3}$ 26. $-\frac{1}{2}e^{-x^2}$ 27. $-\dfrac{e^{-3x}}{3}$

28. $\cos (3-x)$ 29. $\dfrac{\sin^6 x}{6}$ 30. $\log (1+e^x)$

31. 0.312 32. 24.2 33. -0.100

34. 0 35. $16\frac{1}{2}$ 36. 0.693

37. 0.285 38. $\frac{1}{2}$ 39. $\frac{1}{4}$

40. 64

8e   The constant of integration is omitted.

1. $x-3 \log (x+3)$ 2. $\frac{1}{2}x^2 - 3x + 9 \log (x+3)$

3. $\frac{1}{2}x + \frac{3}{4} \log (2x-3)$ 4. $2x + 11 \log (x-4)$

5. $\log (x^2 - 5x + 6)$ 6. $\frac{9}{11} \log (x+6) + \frac{13}{11} \log (x-5)$

7. $x + \log \dfrac{x-2}{x+2}$ 8. $\frac{2}{5} \log (x-1) - \frac{1}{15} \log (3x+2)$

9. $\log (x-1) - \dfrac{2}{x-1}$ 10. $5 \log (x-2) - \dfrac{12}{x-2}$

11. $\log x - \frac{1}{2} \log (2x+3)$ 12. $\log \dfrac{\sqrt{[(x-1)(x-3)^3]}}{x-2}$

13. $\dfrac{1}{2} \log \dfrac{x-1}{x+1}$ 14. $\log \dfrac{\sin x}{1+\sin x}$

15. $\dfrac{1}{12} \log \dfrac{2x-3}{2x+3}$ 16. $\dfrac{1}{4} \log \dfrac{2+x}{2-x}$

17. $\dfrac{1}{2} \log \dfrac{1+e^x}{1-e^x}$ 18. $\dfrac{1}{4} \log \dfrac{1-2 \cos x}{1+2 \cos x}$

19. $\dfrac{1}{2\sqrt{2}} \log \dfrac{x-1-\sqrt{2}}{x-1+\sqrt{2}}$ 20. $\dfrac{1}{4\sqrt{3}} \log \dfrac{x-2-2\sqrt{3}}{x-2+2\sqrt{3}}$

21. $\dfrac{1}{8} \log \dfrac{x-6}{x+2}$ 22. $\dfrac{1}{2\sqrt{5}} \log \dfrac{\sqrt{5}+1+x}{\sqrt{5}-1-x}$

8f   The constant of integration is omitted.

1. $x \log x - x$ 2. $\dfrac{x^2}{2} \log x - \dfrac{x^2}{4}$

3. $-\dfrac{1}{4x^2}(2\log x+1)$

4. $\dfrac{1}{9}(3x\sin 3x+\cos 3x)$

5. $\sin x-x\cos x$

6. $\dfrac{x(1+x)^{11}}{11}-\dfrac{(1+x)^{12}}{132}$

7. $\dfrac{x^8}{64}(8\log x-1)$

8. $\dfrac{x^{\frac{3}{2}}}{9}(6\log x-4)$

9. $e^x(x^2-2x+2)$

10. $(x^2-2)\sin x+2x\cos x$

11. $41\cdot2$

12. $0\cdot330$

13. $7\cdot00$

14. $0\cdot571$

15. $39\cdot1$

8g The constant of integration is omitted.

1. $-\dfrac{\cos 6x}{6}$

2. $-\dfrac{\cos(2x-1)}{2}$

3. $\dfrac{\sin(3x+2)}{3}$

4. $\dfrac{\sin(8t-1)}{8}$

5. $\dfrac{1}{2}\left(x-\dfrac{\sin 6x}{6}\right)$

6. $\frac{1}{2}(t+\sin t)$

7. $-\frac{1}{22}(\cos 11x+11\cos x)$

8. $\frac{1}{8}(\sin 4t+2\sin 2t)$

9. $\frac{1}{8}(2\sin 2t-\sin 4t)$

10. $\frac{1}{12}(3\sin 2x-\sin 6x)$

11. $\frac{1}{42}(7\sin 3x+3\sin 7x)$

12. $\frac{1}{42}(7\cos 3\theta-3\cos 7\theta)$

13. $\frac{1}{4}(2\cos t-\cos 2t)$

14. $-\dfrac{\cos^5\theta}{5}$

15. $\dfrac{\sin^6 t}{6}$

16. $\dfrac{\sin^5\theta}{5}+2\sin\theta$

17. $\dfrac{\cos^3\theta}{3}-\cos\theta$

18. $-\dfrac{1}{2\sin^2 x}$

19. $\log\sec\theta$

20. $\dfrac{\tan^4\theta}{4}$

21. $0\cdot046$

22. $-0\cdot350$

23. $0\cdot4$

24. $0\cdot141$

25. $1\cdot571$

26. $0\cdot786$

Miscellaneous exercises 8

2. (a) i $70\frac{1}{2}$ ii $-0\cdot561$ (b) $3\frac{3}{4}$
3. (a) $8\frac{2}{3}$ (b) $\frac{1}{3}$ (c) $3\cdot195$; $3\frac{1}{3}$
4. (a) i $0\cdot189$ ii $0\cdot390$

5. (a) $\log \sqrt{(1+x^2)} + c$ (b) $\frac{2}{3}x^{\frac{3}{2}} - 4x^{\frac{1}{4}} + c$ (c) $3 \cdot 142$

6. (a) i $\dfrac{2x^2 - 1}{\sqrt{(x^2 - 1)}}$ ii $\dfrac{8e^{4x}}{(e^{4x} + 1)^2}$ (b) i $0 \cdot 693$ ii $3\frac{11}{15}$

7. (a) $\frac{1}{2}$ (b) 8 (c) $0 \cdot 549$ (d) $e^a - e$

8. i $\frac{1}{3}$ ii $0 \cdot 805$ iii $\log 4 - 0 \cdot 75 = 0 \cdot 636$ (b) $10\frac{2}{3}$

9. (a) $c - \log(3 - x)$ (b) i $0 \cdot 500$ ii $14 \cdot 9$ iii $3 \cdot 44$

10. (a) $\frac{1}{2}$ (b) $1 \cdot 386$ (c) $\frac{1}{8}$

11. (a) $\dfrac{3}{x+3} - \dfrac{2}{x+2}$

12. (a) i $9 \cdot 77$ ii $11 \cdot 3$ iii $3 + \frac{3}{4}\pi$
    (b) $3 \log(2x - 3) - \frac{4}{3}\log(3x - 2) + c$

13. (a) $-\frac{1}{3}$ (b) $23 \cdot 7$ (c) $\frac{1}{3}$ (d) $22\frac{2}{3}$

14. (a) i $6\frac{3}{4}$ ii $\frac{1}{2}$

    (b) i $\frac{9}{5}\log(x + 2) - \frac{4}{5}\log(2x - 1) + c$ ii $x^2\left(\dfrac{\log 2x}{2} - \dfrac{1}{4}\right) + c$

15. (b) $c - \frac{1}{3}\log(2 + 3\cos x)$

16. (a) $\dfrac{3}{4}x^{\frac{4}{3}} + \dfrac{3}{x} + c$ (b) $\frac{1}{2}e^{2x-3} + c$ (c) $\frac{2}{3}\sqrt{(2 + 3x)} + c$

    (d) $\dfrac{3\cos 2x - \cos 6x}{12} + c$

17. (a) i $\dfrac{2x(x^2 + 5a^2)}{5\sqrt{x}} + c$ ii $\dfrac{\sin 8x + 4\sin 2x}{16} + c$

    (b) $-2 + \dfrac{1}{1-x} + \dfrac{1}{1+x}; \; 0 \cdot 0878$

18. (a) i $7 \cdot 33$ ii $0 \cdot 865$ iii $5 \cdot 08$ (b) $1 \cdot 26$

19. (a) $0 \cdot 374$ (b) $0 \cdot 707$ (c) $19 \cdot 2$ (d) $c - \dfrac{\cos k\theta}{2k}$

20. (a) i $x\dfrac{e^{2x}}{2} - \dfrac{e^{2x}}{4} + c$ ii $0 \cdot 366$ (b) $1 \cdot 622; 1 \cdot 609$

9d

1. $1\frac{1}{3}$	2. $\frac{1}{6}$	3. 2	4. $2\frac{2}{3}$
5. $\frac{1}{12}$	6. $1\frac{1}{3}$	7. 16	8. $12 \cdot 1\pi$
9. 2π	10. $32\frac{2}{3}\pi$	11. $4\frac{4}{3}\pi$	12. $50\frac{2}{15}\pi$
13. 36π	14. $4\frac{2}{3}\pi$	15. $\frac{2}{3}\pi$	16. 6π
17. $\frac{2}{3}\pi a^3$	18. $\dfrac{\pi a^2 h}{3}$	19. 262	20. (a) $1\frac{1}{15}\pi$
			(b) $8\frac{1}{10}\pi$

21. $(0, 0)(3, 1); \; 1; \frac{9}{10}\pi; 2\frac{7}{10}\pi$ 22. 343 cm^3

9f

1. $2\frac{1}{3}$
2. $1\frac{1}{2}$
3. (a) 0·637 (b) 0
4. (a) $\frac{1}{2}$ (b) $\frac{1}{2}$
5. 1·91
6. 3·91
7. 19 m s^{-1}
9. (a) 40 m s^{-1} (b) $53\frac{1}{3}$ m s^{-1}
10. 250
11. (a) 1·15 (b) 0·730
12. 1·23
13. 3·32
14. 82·5
15. 8·25
16. 4·53
17. (a) 0·85 J
    (b) 10 J
18. 9·6 J
19. 0·5 μJ
20. 11·1 MJ
21. 15·6 kJ, 23·4 kW
22. 672 kJ

9h

1. 405 g, 89·0 cm
2. 15 kg, 3·6 m
3. 1·01 Gg
4. (1·5, 1·2)
5. (1·65, 4·84)
6. (0, 1·6)
7. (1, 0·4)
8. $(\frac{5}{3}, 0)$
9. $(0, \frac{8}{3})$
10. $(\frac{5}{8}, 0)$
11. $(\frac{3}{4}h, 0)$
12. (1·8, 1·8)
13. $\left(\frac{\pi}{4}, \frac{\pi}{8}\right)$
14. $(\frac{3}{8}a, 0)$
15. $(\frac{8}{15}a, 0)$
16. 12·4 cm^2, 1·55 cm^3
17. 3·0 kg
18. 88·5 g
19. 3·53 cm^3, 0·982 g
20. 625 g
21. $\frac{4}{3\pi} \frac{R^2 + Rr + r^2}{R + r}$ (a) $\frac{4R}{3\pi}$ (b) $\frac{2R}{3\pi}$
22. 207 cm^3

9i

1. 0·693 m
2. 10 kg m^2; 0·447 m
3. $\frac{l}{2\sqrt{3}}$
4. 15·6 m^2
5. 0·958 m^2
6. (a) 6·4 (b) $6\frac{2}{21}$; 2·4, $2\frac{2}{7}$
7. 3·6, 0·8
8. $\frac{1}{70}, \frac{3}{10}$
9. $4\frac{4}{9}$, 5·08
10. 0·02, 0·43
11. $\frac{3}{10} Mh^2 \tan^2 \alpha$
12. (a) $\frac{h^2}{2}$ (b) $\frac{h^2}{18}$ (c) $\frac{h^2}{6}$
14. 56 cm^4
15. 396 cm^4, $75\frac{2}{3}$ cm^4

9j

1. 1·25
2. 0·608
3. 37·3
4. 0·956
5. 0·232
6. 3·91
7. 1·18
8. 6·80
9. 1·81
10. 0·847

Miscellaneous exercises 9

1. (a) $(2\frac{1}{2}, \frac{1}{4})$ (b) $\frac{1}{6}$ (c) $\frac{1}{5}$

2. (b) $5\frac{1}{3}, 60{\cdot}3$ (c) $1{\cdot}8$

4. $\dfrac{\pi a^3}{2}$; $(0, \frac{2}{3} a)$

5. (b) 16 (c) $2{\cdot}89$

6. 16, 322; $(0, 3{\cdot}2)$

7. $(\frac{2}{3}, \frac{16}{9}) (2, 0)$; $1\frac{5}{27}$

8. $(5\frac{2}{5}, 3\frac{3}{8})$, 1150

9. $1, 1\frac{1}{4}, -\frac{7}{20}$

10. (a) $6{\cdot}64, 0{\cdot}916$

11. (a) $0{\cdot}0471$ (b) ii $18{\cdot}3$

12. (a) i $1\frac{3}{4}$ ii $3{\cdot}76$ (b) $k \log \dfrac{v_2 - b}{v_1 - b} - \dfrac{a(v_2 - v_1)}{v_1 v_2}$

13. (a) $10, 0$ (b) $16{\cdot}8$

15. (b) 34167 m^4

16. (a) $1{\cdot}5, 480, 720$

17. (a) $7\frac{3}{5}, 3\frac{4}{5}$ (b) $0{\cdot}983$

18. $\dfrac{bd^3}{3}, 44\frac{1}{6}$

19. 2890 cm^3

20. (b) 334 cm^4

21. (b) 189 cm^2

22. (b) $203{\cdot}7 \text{ cm}^4, 35{\cdot}3 \text{ cm}^4$

23. (c) $976 \text{ cm}^4, 200 \text{ cm}^4, 1176 \text{ cm}^4$

24. (b) $\frac{5}{3} \text{ cm}^4, 21{\cdot}35 \text{ cm}^4, 23{\cdot}0 \text{ cm}^4$

10a

1. $y = x\dfrac{dy}{dx} + 1$

2. $x\dfrac{dy}{dx} = 2y$

3. $4y = x\dfrac{dy}{dx}$

4. $\dfrac{d^2y}{dx^2} = \dfrac{1}{x}\dfrac{dy}{dx}$

5. $y = x\dfrac{dy}{dx} + \dfrac{2}{x}$

6. $y = x\dfrac{dy}{dx} - 2x^3$

7. $x^2\dfrac{d^2y}{dx^2} - 3x\dfrac{dy}{dx} + 3y = 0$

8. $\dfrac{d^2y}{dx^2} - \dfrac{4}{x}\dfrac{dy}{dx} + \dfrac{6}{x^2} = 0$

9. (a) $y = \dfrac{5x^2}{2} + 2x + c$ (b) $y = x^2 + \dfrac{x^3}{3} + c$

(c) $y = \sin 2x + c$ (d) $y = 2x^3 + \dfrac{3}{x} + c$

(e) $y = e^x + c$ (f) $y = \dfrac{x^2}{2} + \log x + c$

10. $y = x^4 - x^2 + 2$

11. (a) $y = x^3 - 3x^2 + 2$ (b) $y = 3x + \dfrac{1}{x^2}$

(c) $y = \dfrac{e^{2x} + e^{-2x}}{2} + 1$

12. (a) $s = 10t - t^2 + 24$ (b) $p = \frac{5}{2}v^2 - 2v + 5\frac{1}{2}$

    (c) $s = 2t + \dfrac{t^2}{5} - 3$

10b

1. $y = e^x + x^2 + c$

2. $y = \log x - \dfrac{x^2}{2} + c$

3. $y = \frac{1}{2}(\sin 2x - \cos 2x) + c$

4. $y = \log x - \dfrac{2}{x} + c$

5. $y = c - \dfrac{2}{x} - \dfrac{3}{2x^2}$

6. $y = \dfrac{3}{2}x + \sin x + \dfrac{\sin 2x}{4} + c$

7. $y^2 = x^2 + 2\log x + c$

8. $y + 3 = c(x + 2)$

9. $y = Ae^{-2x} + \frac{3}{2}$

10. $y = Ce^{x^2 - x}$

11. $x + y + c = \log(x + 1)(y + 1)$

12. $y = Ce^x$

13. $v = \dfrac{g}{k} - Ce^{-kt}$

14. $v^2 = 2x^2 - 8x + c$

15. $N = N_0 e^{kt}$

16. $N = N_0 e^{-kt}$

17. $y = 4e^{-x} + Ax + B$

18. $y = \dfrac{x^4}{2} + \dfrac{\sin 2x}{4} + Ax + B$

19. $s = t \log t + At + B$

20. $x = At + B - \cos t$

21. $y = 2e^{-x}$

22. $y = \log x - \dfrac{1}{x} + 1$

23. $\log x = \log t + t - 1$

24. $r = 3e^{\cos \theta}$

25. $\theta = 60e^{-kt}; k = 0.00105$

26. $i = \dfrac{E}{R}(1 - e^{-Rt/L})$

27. $\omega = \omega_0 - \dfrac{Nt}{I}, 471$ s

28. 0.000167

10d

1. $y = 5\sin 2x, 3.14$

2. $s = 2\sin t, 6.28$

3. $\theta = \theta_0 \cos \sqrt{\left(\dfrac{g}{l}\right)}t, 2\pi\sqrt{\dfrac{l}{g}}$

4. $x = a\cos \sqrt{\left(\dfrac{s}{m}\right)}t, T = 2\pi\sqrt{\dfrac{m}{s}}$

5. $\theta = 4\sqrt{\dfrac{Il}{JCg}} \sin \sqrt{\left(\dfrac{JCg}{Il}\right)}t; T = 2\pi\sqrt{\dfrac{Il}{JCg}}$

6. $y = 4\sqrt{2} \sin\left(\dfrac{x}{2} + \dfrac{\pi}{4}\right), T = 4\pi = 12.57$

7. (a) $a = 2, b = 0, n = 6$ (b) $a = 0, b = 10, n = \frac{1}{2}$

8. 0.448 s, 2.23

9. $\dfrac{1}{2\pi}\sqrt{\dfrac{3EI}{Ml^3}}$

10. 04.05 h, 0.029

Miscellaneous exercises 10

1. (b) 1, 2

3. $2x\dfrac{dy}{dx} = y$ (a) $y = x^2 + 4\log x + c$ (b) $N = 20(2 - e^{-\frac{1}{2}t})$

4. (a) i -1 ii 0.223 iii $\frac{8}{15}$ (b) $x = \dfrac{\cos 2t}{2} - \dfrac{\sin 2t}{4}$

5. (a) $1, -\frac{1}{2}, \frac{1}{12}$ (b) $y = e^x - e^{-x} - x + 3$

7. (a) i $\dfrac{3}{x}$ ii $\dfrac{2\cos x}{(1+\sin x)^2}$ iii $e^{\frac{1}{2}x}(\frac{1}{2}\tan x + \sec^2 x)$

   (b) $y = 2\log x + x - 1$

8. (b) $s = t^3 + \dfrac{9}{2}t^2 - 6t + 12$ (c) $y = x^2 - \dfrac{x^3}{9} - 2$

9. (a) i $-\dfrac{1}{x}$ ii $-\dfrac{1}{2}\cos\left(\dfrac{\pi}{6} - \dfrac{x}{2}\right)$ iii $\dfrac{2x}{(x^2+1)^2}$

10. (a) $1 - \dfrac{x}{2} + \dfrac{x^2}{8} - \dfrac{x^3}{48}$ i 0.861 ii 0.019 (b) $2, 0.927$

11. (a) $x\dfrac{d^2y}{dx^2} = \dfrac{dy}{dx}$ (b) i $y = e^x + 2\cos x + c$ ii $y - 2 = c\sqrt{(2x-1)}$

12. (a) $y^2 = 2\log x + 4x - x^2 + c$ (b) $x = 4e^{-t} + At + B$

   (c) $y = 2 - \dfrac{2}{x} - 4\log x$

13. $x = \dfrac{v_0}{n}\sin nt$, 31.4 m s^{-1}, 1970 m s^{-2}

14. $x = 10\cos 3t$, 0.477

15. (a) $\dfrac{dy}{dx} = \dfrac{xy}{x^2-1}$ (b) i $T = T_0 e^{\mu\theta}$ ii $y + 2 = Ae^{\frac{1}{2}x^2}$

16. (b) i $y^2 + 1 = Ae^x$ ii $EIy = \dfrac{w(l-x)^3}{6} + A(l-x) + B$

17. (a) $y = 4\log\sec x + c$ ii $i = \dfrac{E}{R}(1 - e^{-Rt/L})$

   (c) $v^2 = 100 - 2gx$

18. 62.8 m s^{-1}, 1970 m s^{-2}

19. $s = 2\sin 4t$, 6.93 m s^{-1}, 16 m s^{-2}

20. (a) $y = 2x^2 + e^{-x} + c$ (b) $s = t^3 - 2t^2 + 4t + 10$

   (c) $\sin y = \sin x + \dfrac{e^{2x}}{2} + c$

11a 1. $\bar{x} = 6.5$, $\sigma = 1.7$ 2. 31.42 kN, 0.679 kN
     3. $4; 8, 2$ 4. $3.74, 18$

5. 3291, 26

6. 43 kg, 3 kg

7. A 1·992, 0·0308   B 2·003, 0·0475

8. 5·292 cm, 0·041 cm

11b

1. (a) 27·9 kg, 0·538 kg   (b) 3308·3 m, 63·3 m
   (c) 275·9, 0·6245
2. 54·25 m, 6·85 m

3. 1236·4 m, 9·33 m

4. 67·38 cm, 2·36 cm, 2·34 cm, 3·47
5. 259 m, 9·6 m

6. 42·7 cm, 1·2 cm

7. 25·5 kN, 0·257 kN
8. 32·85, 0·342
9. 7·825, 0·818

10. 67·8 cm, 2·90 cm

11e

1. 49·81, 50
2. 32·66, 32·88, 33·13, 0·235, kN
3. 98·38, 101·42, 1·52, cm
4. 99 cm, 99·85 cm, 99·88 cm
5. 29 cm, 28·5 cm, 1·86 cm
6. 32 cm, 33·1 cm, 33·6 cm
7. 0·630, 0·909, 0·140
8. 0·99
9. 169, 151·3, 190·1
10. 0·9971 cm, 0·9981 cm, 0·9960 cm

11f

1. (a) $\frac{1}{13}$ (b) $\frac{1}{52}$
2. (a) $\frac{1}{6}$ (b) $\frac{5}{6}$ (c) $\frac{1}{3}$ (d) $\frac{1}{36}$ (e) $\frac{5}{36}$ (f) $\frac{1}{12}$ (g) 1
3. (a) $\frac{1}{32}$ (b) $\frac{5}{16}$ (c) $\frac{5}{16}$

4. (a) $\frac{3}{5}$ (b) $\frac{2}{5}$

5. (a) $\frac{7}{22}$ (b) $\frac{49}{144}$

6. (a) $\frac{21}{55}$ (b) $\frac{27}{64}$

7. (a) $\frac{1}{15}$ (b) $\frac{7}{15}$ (c) $\frac{7}{15}$
8. (a) $\frac{27}{64}$ (b) $\frac{27}{64}$ (c) $\frac{9}{64}$ (d) $\frac{1}{64}$
9. $\frac{1}{16}, \frac{4}{16}, \frac{6}{16}, \frac{4}{16}, \frac{1}{16}$;

x	0	1	2	3	4
f	8	32	48	32	8

10. (a) $\frac{32}{243}$ (b) $\frac{80}{243}$ (c) $\frac{1}{243}$ (d) 1   (e) 32
11. 0·085

12. (a) 0·294   (b) 0·046

11h

1. (a) 0·6826   (b) 0·7745   (c) 0·9973
2. (a) 0·7745   (b) 0·6730
3. 4

4. (a) 15·9%   (b) 93·3%

5. 0·533

6. (a) 0·667   (b) 0·670

7. 0·301, 0·361, 0·217, 0·087

8. 1·900 and 2·116 cm   *9.* 10 hr
10. 1·234 and 1·290 N, 1·244 and 1·280 N

Miscellaneous exercises 11

1. 119·4, 1·6

2. 5542, 147

3. A 323, 1·095; B 324, 1·342; A

4. 4·86, 5, 5

5. (b) iv 3·13   v 3·22   vi 3·39

6. 0·178 48, 0·000 139

7. 7·20, 7·82, 8·44, 0·62, kN

8. (b) 7·609 mm, 0·002 mm

9. 59 hr, 10 hr

10. $\frac{1}{32}, \frac{5}{32}, \frac{10}{32}, \frac{10}{32}, \frac{5}{32}, \frac{1}{32}$;

x	0	1	2	3	4	5
f	8	40	80	80	40	8

11. (a) $\frac{1}{15}, \frac{7}{15}$ (b) $\frac{9}{100}, \frac{49}{100}$

12. 2552 h, 96·6 h

13. (a) 4008   (b) 20478

14. 0·2231, 0·3347, 0·2510, 0·1255

15. 20·40, 1·40, kN; 6·86

16. (a) 0·7745   (b) 6, 19

12c

3. $1 + j8$

4. $3 + j4$

5. $-j3$

6. -3

7. $j3$

8. $-48 + j44$

9. $4 + j88$

10. -14

11. 0

12. j

13. $\frac{19}{13} - j\frac{9}{13}$

14. $0·432 - j0·763$

15. $-0·15 - j0·65$

16. (a) $\frac{3}{13}, -\frac{2}{13}$ (b) $-0·32, 0·24$ (c) 14, 4

17. (a) $11 + j27$ (b) $-8·2 + j0·6$ (c) $-\frac{1}{2} + j\frac{\sqrt{3}}{2}$

18. (a) $\frac{19}{13} - j\frac{9}{13}$ (b) $-1 \pm j\sqrt{5}$

19. (a) $4 - j3$ (b) $-4 + j3$

20. $5 + j2, 1$

12g

1. $5\underline{/53° 8'}$

2. $5\underline{/\pi}$

3. $5\underline{/143° 8'}$

4. $7·07\underline{/81·80°}$

5. $4\underline{/-\frac{\pi}{2}}$

6. $9·45\underline{/122°}$

7. $21·1\underline{/-95·5°}$

8. $1\underline{/-43° 36'}$

9. $17·5\underline{/-76° 44'}$

10. $10\underline{/0°}$

11. $3·46 + j2$

12. $2·70 - j3·22$

13. $0·677 + j2·717$

14. $j10$

15. -6

16. $-4·10 + j4·10$

17. $15\underline{/45°}$

18. $17·0 \underline{/ 121°}$

19. $4\underline{/20°}$

20. $0·5\underline{/53°}$

21. $5·83\underline{/82°}$

22. $0·9\underline{/157°}$

23. $\sqrt{13}\underline{/33·7°}$; $610\underline{/169°}$

24. $676\underline{/45\cdot2°}$

25. (a) $1, -\theta$ (b) $2, -\dfrac{\pi}{4}$

   (c) $15\cdot1, -172\cdot4°$

27. (a) $0\cdot404+j6\cdot43, 6\cdot45\underline{/86°\ 24'}$ (b) $3\cdot06-j4\cdot43, 5\cdot38\underline{/-55°\ 18'}$

28. (a) i $3\cdot61, 33\cdot7°$ ii $23\cdot6, 36°\ 23'$

   (b) $-1\pm j4; \sqrt{17}\underline{/\pm104°}$ (c) $-\tfrac{3}{25}, -\tfrac{4}{25}$

29. $13\cdot4\underline{/-53\cdot7°}$ 30. $13\cdot1$ amp, $17\cdot6°$ lead

Miscellaneous exercises 12

1. (a) $13, 22°\ 37'$ (b) $-0\cdot32j$
2. (a) i $200-150j$ ii $1+j$
3. (a) $\pm j2$ (b) $\dfrac{1}{2}\pm j\dfrac{\sqrt{3}}{2}$
4. (a) $340+j80$ (b) $19\cdot0+j2\cdot45$
5. (a) $1-j$ (b) $3+4j$
6. (a) $-\dfrac{1}{2}\pm j\dfrac{\sqrt{7}}{2}, \sqrt{2}\underline{/\pm110\cdot7°}$ (b) i $4\cdot59-j0\cdot74$ ii $20\cdot8-j5\cdot6$
7. (a) $-3\pm j4$ (b) $39+j25$ (c) $4+j$
8. $a=3, b=4; -2+j$
9. $13\cdot2\underline{/-7\cdot6°}$ amp   10. $31\cdot5\underline{/-20\cdot3°}$ amp
11. $128\underline{/55°\ 12'}$ 12. $4\cdot47\underline{/24°\ 15'}$ (b) $1\cdot07\underline{/46°\ 56'}$

Examination paper 1

1. (a) $-\tfrac{1}{3}, \tfrac{1}{3}, -1$ (b) $x=-2$ (c) $\dfrac{2}{x-2}-\dfrac{1}{x+1}+\dfrac{3}{(x+1)^2}$

2. (a) $1\cdot0516$ (b) -3%

3. (a) $13\sin(\omega t-1\cdot176); 13; \dfrac{2\cdot75}{\omega}$

   (b) $48°\ 35', 131°\ 25', 270°$

4. (a) $P=1\cdot635, W=1\cdot266$ (b) $0\cdot41421, -2\cdot41421$

5. (a) i $e^{2x}(3\cos3x+2\sin3x)$ ii $\dfrac{2}{x^2(2x+1)}-\dfrac{2\log(2x+1)}{x^3}$

   iii $\dfrac{x}{\sqrt{(x^2-1)}}$

   (b) max. -2; min. $+2$

6. (a) i $\dfrac{x^3}{3}+\dfrac{1}{x}+c$ ii $6e^{\frac{1}{6}x}+c$ iii 1 (b) $3\tfrac{3}{4}, 3\cdot7743$

7. (b) 13 m s^{-2}, $388\cdot4$ m

8. (a) $\dfrac{R_2^2+R_1^2}{2}$ (b) i $\dfrac{\pi}{4}(R_2^4-R_1^4)$ ii $\dfrac{\pi}{4}(5R_2^4-4R_1^2R_2^2-R_1^4)$

9. $32 \cdot 89, 0 \cdot 368$

10. (a) $x^2 - \dfrac{x^3}{3} + \dfrac{1}{6}x^4$ (b) 7%

Examination paper 2

1. (a) $1 + 2x^2 + 3x^4 + 4x^6$; $1 \cdot 0080$ (b) $2 + \dfrac{x^2}{8} + \dfrac{x^4}{192}$ (c) $5\frac{1}{2}\%$ decrease
2. (b) $R = 8 \cdot 06, \alpha = 60° \ 15'$; $8 \cdot 06, 150° \ 15'$
3. (a) i $\dfrac{1 + 2x - x^2}{1 - 2x + x^2}$ ii $3 \cos 3x \cos 2x - 2 \sin 3x \sin 2x$ iii $\dfrac{x}{x^2 - 3}$
   (b) $0 \cdot 32$ cm s^{-1}; $0 \cdot 051$ cm s^{-1}
4. (a) $X = x^2, Y = y$; $X = x, Y = \dfrac{1}{y}$; $X = x, Y = \log y$
   (b) $\log (4c) \times$ linear; $A = 4 \cdot 8, n = 6 \cdot 9$
5. $4 \cdot 991, 4 \cdot 990, 5 \cdot 000, 4 \cdot 982, 0 \cdot 009$ (cm)
6. (a) $4 \cdot 5826$ (b) $4 \cdot 343$
7. (a) i $2x^{\frac{3}{2}} + \dfrac{3}{x} + c$ ii $\dfrac{\sin^3 x}{3} + c$ (b) i $4 \cdot 57$ ii $0 \cdot 112$
8. $6 \cdot 56, 84 \cdot 8, 2 \cdot 06$
9. (a) $2 \cos (2x + 3y), 3 \cos (2x + 3y)$ (b) $1 \cdot 27$
10. (a) $y - y_1 = m(x - x_1), 2x + y = 10$ (b) $y + 1 = 2\sqrt{(2x - 1)}$

Examination paper 3

1. (a) $0 \cdot 6931, -0 \cdot 9163$ (b) $\frac{1}{2}, -\frac{1}{2}, -\frac{3}{4}$
   (c) $\dfrac{17}{12(x + 2)} + \dfrac{1}{3(x - 1)} - \dfrac{3}{4(x - 2)}$
2. (a) $1 \cdot 8$ (b) $19° \ 28', 160° \ 32', 210°, 330°$
3. (a) i $1 + \dfrac{x}{3} - \dfrac{x^2}{9} + \dfrac{5}{81}x^3$ ii $1 + \dfrac{2}{3}x + \dfrac{8}{9}x^2 + \dfrac{112}{81}x^3$ iii $1 + x + x^2 + \dfrac{5}{3}x^3$
   (b) $25(2x + 4x^2 + 8\frac{2}{3}x^3 + 20x^4)$
4. (a) i $e^{2x}(2x^2 - 4x - 3)$ ii $\dfrac{1 - x^2}{(1 + x^2)^2}$ iii $4 \tan (2x - 1) \sec^2 (2x - 1)$
   (b) $(1, 6 \cdot 5)(-2, 0)$
5. (a) $-\dfrac{1}{2} + j\dfrac{\sqrt{11}}{2}, -\dfrac{1}{2} - j\dfrac{\sqrt{11}}{2}$ (b) $0 \cdot 4 - j7 \cdot 8$ (c) $15 \cdot 4 \underline{/-76\frac{1}{2}°}$
6. (a) $3 \cdot 461$ (b) $1 \cdot 58414$
7. (a) i $\dfrac{x^2}{2} - 2x + \log x + c$ ii $c - \dfrac{1}{3(2 + 3x)}$
   (b) $-0 \cdot 561$ (c) $2 \cdot 83$
8. $(1 \cdot 846, 0)$
9. $x = 12 \cos 4t$; $12, 0 \cdot 637$
10. (a) $0 \cdot 7745$ (b) $13; 62$

Index